FOREWORD

In 2012, I began to document the 1,700-acre Douglas Missile Test Facility for the Historic American Engineering Record (HAER), a heritage program administered by the National Park Service intended to document historic sites and structures related to engineering and industry. This site, so important to aeronautical history, had largely been abandoned for more than three decades and was slated for demolition to make way for the development in Rancho Cordova—a growing suburb of Sacramento.

As an architectural historian and historical archaeologist, I had previously recorded military and engineering facilities that postdated World War II, but I wanted to know more about aerospace history. I needed to find an expert. Internet searches on the topic led me to Alan Lawrie. He published an article on the Douglas facility and knew so very much about the Saturn program and its engineering. I also enlisted the expertise of Robert Hicks, a photographer with the skills to meet the exacting standards of the HAER program.

Long rows of piled high cobbles and silt, interspersed with deep rows of excavated earth, provided a setting that was spacious enough to build a large-scale facility and simultaneously provide a buffer for captive firings and any explosions that would—and did—occur. It was also a landscape situated near the urban population of Sacramento, which would provide the necessary work force.

Several complexes (Alpha, Beta, Gamma, Kappa, Sigma, Solid Propellant, and the administrative area) make up the Douglas Missile Test Facility, each serving a different function. The site was associated with the historical engineering development of the Nike Hercules surface-to-air missile, Thor intermediate-range ballistic missile, Nike Zeus (an anti-missile missile), Titan LR-91 second stage engine, Skybolt interceptor missile, and the NASA Saturn S-IVB booster, as well as the development and testing of liquid hydrogen, liquid oxygen, and bipropellant Auxiliary Propulsion Modules. These complexes contain specialized buildings and structures in a compatible landscape that reflects technological advances in form, function, and historic aesthetic. Each complex, and the activities associated with it, have left physical traces on the landscape. The individual complexes also relate to one another across time and space to create a whole. Overall, the facility captures the development and testing of the rocket booster engines associated with the nation's military innovations and space exploration.

To adequately capture the depth, breadth, and context of this unique facility, we needed architectural, historical, engineering, and archaeological perspectives. The HAER record attempts to capture the study of an evolution of technology and the historical and physical traces that it left behind. The photographic and documentary HAER record (CA-HAER-2310) is now at the Library of Congress and can be found online at loc.gov.

—Rebecca Allen
Cultural Resources Director
Environmental Science Associates

INTRODUCTION

In 1849, gold was discovered in the foothills east of Sacramento, on the American River. Much of this gold was in flake form, so it permeated the rocky soil to a depth of 100 feet. In the early 1900s, a gold mining firm (the Natomas Company) started commercial dredging. In this process, a large barge was floated on a man-made lake. Buckets would bring a slurry of rock, soil, gold, and water to the barge, where it was floated across a large table filled with mercury. The flake gold would amalgamate with the mercury, and the remaining slurry would wash out and be dumped onto the backside of the lake. This resulted in the land being literally turned upside down. Vast areas of land were left unusable for farming or home construction.

In the early 1950s, Aerojet, a spinoff from Caltech's Jet Propulsion Labs in Pasadena, California, needed a much larger facility to develop rocket propulsion systems. It selected Sacramento for a number of reasons, particularly the abundance of inexpensive and otherwise unusable land and the availability of the necessary support infrastructure. Aerojet proceeded to purchase about 26,000 acres. A vast manufacturing and test complex was constructed for both solid and liquid propulsion systems. In 1955, the Douglas Aircraft Company was awarded an Air Force contract to design, develop, and deliver the Thor intermediate-range ballistic missile (IRBM). Douglas needed to construct a test facility for this and negotiated to purchase 2,000 acres from Aerojet, with an option for 2,000 more. In 1955, the construction of the new Douglas Sacramento Test Center (STC) began.

By mid-1957, the Alpha complex was completed, and testing of the Thor missile started. This was an exciting and difficult learning experience for us young engineers. We were developing a large missile with a powerful liquid-fuelled engine. The propellants (RP-1 kerosene and liquid oxygen [LOX]) had to be stored and then transferred to the missile in the test stand. LOX was used at minus 296 degrees Fahrenheit, which affected all of the components used, and the vibration added more stress to the components. We would start a countdown to static fire the missile in the morning. Each subsystem would be manually functionally tested, data would be reviewed, and then the missile would be loaded with propellants. Countdowns would last for hours—usually delayed by some components developing problems. The final countdown lasted about 15 minutes and involved topping off the missile propellant tanks and pressurizing all the gas bottles and missile propellant tanks to required start pressures. All the while, a number of engineers were sitting at their subsystems control panels and monitoring every important parameter. Any variant in data would cause a hold to occur, followed by troubleshooting. A trouble-free countdown was scheduled at three to four hours. In the early days, it was not uncommon to be in countdown for 12 to 16 hours. Many of us enjoyed the thrill of sleeping on an office chair or lying against a control panel (some of the toggle switches would leave a lasting indent in your cheek!). As time went on, the components and systems matured, and testing proceeded at a more scheduled pace.

Testing a large missile with these propellants was hazardous. We took every precaution known at that time to ensure the safety of the personnel and the equipment. But, since we were pioneering new ground, there were things we had to learn the hard way. One of the test stands was called the Initial Operational Capability (IOC) Site. This was a design for the tactical launch site to be built in Europe, where the Thor IRBM was to be deployed. The design of this site kept the Thor IRBM laid out horizontally on an erector device, covered by a moveable garage structure. Fuel and oxidizer propellant storage tanks were nearby and connected to the missile. The tactical countdown to launch was planned at 15 minutes from "GO." In that time

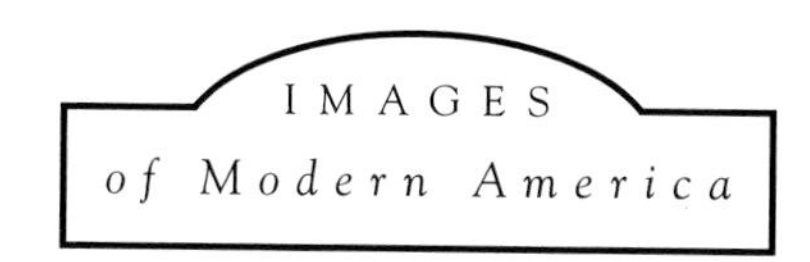

SACRAMENTO'S MOON ROCKETS

This photograph shows the test firing of the Saturn V third stage, which took the crew of Apollo 11 to the moon for the historic first landing in July 1969. The S-IVB-506N stage was test fired for 445.2 seconds at Douglas Aircraft Company's Sacramento testing facility's (SACTO) Beta III test stand on July 17, 1968. (68-68410. Alan Lawrie/Arlene Royer/National Archives and Records Administration [NARA].)

On the Front Cover: Clockwise from top left:
The S-IV Battleship undergoes a test firing (Gene Robinson; page 27); the entrance to the Beta complex (Vince Wheelock; page 44); NASA and Douglas employees donating children's gifts to the Salvation Army (Julian Pennello; page 53); the storage, modification, and checkout of the S-IVB Auxiliary Propulsion System modules (Phil Broad/Mike Jetzer; page 54); technicians in the clean room at SACTO (Dick Serrano; page 53).

On the Back Cover: Left to right:
The air-conditioned control room in the Beta control center (Dick Serrano; page 52); the S-IVB-206 stage being hoisted onto the Beta III test stand (Arlene Royer/NARA; page 45); an artist's rendering of the Beta test complex (Ralph Allen/Marshall Space Flight Center [MSFC]; page 16).

Sacramento's Moon Rockets

Alan Lawrie
Foreword by Rebecca Allen
and Introduction by Don Brincka

ISBN 978-1-4671-3389-0

Published by Arcadia Publishing
Charleston, South Carolina

Printed in the United States of America

Library of Congress Control Number: 2015943994

For all general information, please contact Arcadia Publishing:
Telephone 843-853-2070
Fax 843-853-0044
E-mail sales@arcadiapublishing.com
For customer service and orders:
Toll-Free 1-888-313-2665

Visit us on the Internet at www.arcadiapublishing.com

This book is dedicated to two of my great uncles who gave their lives in the Great War. Alexander Lawrie (left) was born June 24, 1891, at 21 Rossie Place, Edinburgh. He served in the 16th Battalion Royal Scots and was killed in the Battle of the Somme on July 1, 1916. John Lawrie (right) was born on February 4, 1896, at 21 Rossie Place, Edinburgh. He served in the 2nd/4th Battalion York and Lancaster Regiment and was killed near Rheims on July 27, 1918. (Both, Anne McLaren.)

Contents

Acknowledgments		6
Foreword		7
Introduction		8
1.	Building SACTO	11
2.	Early Testing	23
3.	Transportation	31
4.	S-IVB Testing	43
5.	Explosions	69
6.	50 Years Later	79
Bibliography		95

ACKNOWLEDGMENTS

These photographs of Douglas Aircraft Company's Sacramento test facility (SACTO) came from a number of sources. The caption to each photograph identifies the source, but I provide more background here. I collected around 900 SACTO photographs and have selected around 160 for this book, based on criteria of color, quality, and subject matter.

I gratefully acknowledge the support of the following people who contributed to this book. Don Brincka showed me around SACTO in 2006, provided his personal photographs and documents, answered numerous questions, and wrote the introduction. Rebecca Allen led the 2013 site survey of SACTO, invited me to participate, and wrote the foreword. Rebecca was supported by Katherine Anderson, Scott Baxter, and Lisa Westwood of Environmental Science Associates and Robert Hicks and Star AndersonHicks.

Terri Pennello (whose father, Julian, worked for NASA at SACTO from 1964 to 1970) provided frequent advice and coordination with the SACTO retirees. Terri also championed the preservation of over 400 photographs stored for 40 years in the garage of retiree Jim Porter and now in the Sacramento Center for History. Ralph Allen, NASA Marshall Space Flight Center (MSFC) retired, provided photographs from the MSFC archives. Arlene Royer, of the National Archives and Records Administration (NARA) in Atlanta, scanned and provided photographs from the original negatives held at NARA.

Mike Jetzer (of heroicrelics.org) supplied various diagrams and made the DVD video capture images of the SACTO explosions. In addition, Mike is now the host of the collection of Douglas photographs—previously located on Phil Broad's Cloudster website—which originally came from the Santa Monica Museum of Flight.

Douglas retirees Val Sushkoff, David Dallman, Dick Serrano, Jerry Robison, and Gene Robinson and Rocketdyne retiree Vince Wheelock provided photographs from their personal collections. Steve Seabourne, president of Automotive Importing Manufacturing, Inc., and current owner of the Vehicle Checkout Laboratory (VCL) provided access to the VCL. Aerojet Rocketdyne, Elliot Homes, and Easton Development Company, LLC, provided access to the site in 2013. Jeff Ruetsche and Ginny Rasmussen kindly and successfully managed the book at Arcadia Publishing.

My thanks to everyone who worked for Douglas and NASA in the 1950s and 1960s, who made this extraordinary story happen. As always, my thanks to Olwyn Lawrie, for everything.

—Alan Lawrie
Hitchin, England
February 2015

frame, we had to roll back the garage structure, erect the missile, load propellants, charge all the gas spheres, run an electrical checkout, program in the target selection, and then launch the missile. In those early days of using liquid oxygen, we were concerned about using gaseous nitrogen (at ambient temperatures) to pressurize the liquid oxygen in the storage tanks and transfer the liquid oxygen into the missile. We were concerned that the gaseous nitrogen would diffuse into the liquid oxygen and degrade rocket engine performance. We opted to use gaseous oxygen to pressurize storage tanks. We designed a gaseous oxygen pressurization system at 3,000 psig, with two-stage regulation, to pressurize the LOX storage tank to 35 psig for propellant transfer. We used commercial stainless steel regulators for this task. One day, we were testing the tactical launch sequence. There were 10 of us at the LOX valve complex setting up the system and verifying operation. At the "GO" signal, the 3,000 psig gaseous oxygen tank shut-off valve was commanded open. The high-pressure oxygen slammed into the first regulator, and the regulator exploded—showering us with flame, molten metal, and pure oxygen gas. The pure oxygen gas saturated our clothing and the flames ignited it. We all received burns—the severity of which depended on how close to the failed regulator we were. Two men died, four spent up to a year in hospital, and four others—myself included—received lighter degrees of burns and required brief hospitalization. After the accident, we embarked on a test program to determine how detrimental using gaseous nitrogen to pressurize LOX was to engine performance. Several months of testing determined it to be acceptable.

Douglas was awarded the contract to build the S-IV stage of the Saturn launch vehicle for the Apollo moon program. This was a challenge, as we were not familiar with liquid hydrogen (LH2) handling and usage. First and foremost, we had to develop a flight vehicle insulation system to maintain LH2 in the flight tankage. Ground storage vessels are all vacuum jacketed, but this was not feasible for a flight vehicle. The industry concept, at that time, was that hydrogen was very dangerous! It was said that it could leak out of the tightest joint and easily ignite, and that the flame was colorless so not easily detected. While we were designing the modifications to the Alpha 1 and 2 test stands for testing the vehicle, we continued testing materials for insulating the vehicle LH2 tank. Douglas engineers decided the best engineering approach to insulating the flight vehicle LH2 tank was for an internally applied material. This ensured that the glue bond where the material is affixed to the flight skin would be at or close to ambient temperature, improving the success of the adhesive. We developed an insulation material made of Styrofoam, with nylon threads that bonded to the inside of the vehicle hydrogen tank very successfully. The S-IV stage, designed by our engineers at the Santa Monica plant, was the largest LH2/LOX stage (18 feet in diameter) at that time. It was to be propelled by four newly designed 15,000-pound-thrust LH2/LOX Pratt & Whitney engines. The flight vehicle LOX tank was essentially a sphere beneath the flight vehicle LH2 tank joined by a common bulkhead.

We were directed by NASA to increase the stage's thrust by increasing the number of engines from four to six. This meant modifications to the stage design and the test stand design. The Pratt & Whitney engines were designed for altitude operation. This meant the engines could not be static fired at sea level conditions without destroying the expansion nozzle. We had to design a system that provided simulated altitude pressures around the expansion nozzle. This was accomplished by surrounding each expansion nozzle with a chamber approximately 30 feet long in the shape of a DeLaval nozzle. The chamber was drawn down to extremely low pressure by a steam ejector system. When the pressure was low enough, the engines would be fired and blow off doors at the bottom of the DeLaval nozzle chamber. This system worked well.

In 1964, we installed a specially designed test vehicle, the All Systems Vehicle, in test stand Alpha 1. It was to be used to explore various test conditions and define operational limits. During a static firing countdown, a problem occurred—delaying the normal progress to firing the engines. The vehicle tanks were pressurized when the vehicle LOX tank's vent and relief valve froze shut. Pressure built up in the LOX tank and wasn't relieved. The LOX tank ruptured, driving the common bulkhead up into the LH2 tank. The stage exploded, completely destroying it. Test stand damage was surprisingly small.

About that time, NASA awarded Douglas the contract to design, build, and test a new stage—the S-IVB—for the Saturn V launch vehicle, which would take our astronauts to the moon and return them safely. The S-IVB stage was significantly larger than the S-IV stage. This required that new test facilities be constructed. Douglas exercised its option and purchased the adjoining 2,000 acres. At NASA's direction, the facility was designed for three test stands, Beta I, Beta II, and Beta III. Construction began on test stands Beta I and Beta III, while construction of Beta II was held back until launch frequency validated the need for a third test stand. This new facility was to be a significant advancement in technology from all of the existing launch and test facilities. This new plan also allowed that the vehicle checkout, test, and launching would be accomplished by a computer.

A large computer, featuring rows of one-inch tape decks, dominated the control room. Subsystem checkout, propellant loading, countdown, and static firing all would be performed by this machine. Up to then, the engineers who wrote the procedures would manually perform the checkouts and test. It was mind-boggling to us test engineers to think of turning over control of a flight stage to an inert object like the computer. The test commands were progressively loaded into the limited 16 kilobytes of RAM in the computer. We were extremely cautious when we started—starting with "single step" commands where we allowed the computer to send one command and stop. We did this multiple times for every checkout procedure. When we were comfortable that the program was correct, we started running total subsystem checkouts. The actual static firing countdown was carefully programmed with key stop points so we could be assured that all systems were operating as required. In a reasonably short time, we demonstrated that the entire countdown could be performed by the computer system. Afterwards, all of the flight vehicle acceptance tests were conducted by the computer system.

In January 1967, we were in a static firing countdown of stage S-IVB-503 on test stand Beta III. Countdown progressed normally until moments before ignition, when the stage exploded. Pandemonium broke out in the control room when the large boom, and loss of all visual and data communication with the test stand, occurred. The flight stage was totally destroyed, and the test stand suffered damage. What followed was weeks of exhaustive data analysis and debris inspection, which determined that the explosion was caused by a high-pressure gas sphere (3,000 psig) splitting apart at its weld seam. This drove the upper hemisphere up through the stage's LOX tank and then the LH2 tank. The stage was fully loaded with propellants and pressurized. Subsequent investigations found that the supplier of the gas spheres had used the wrong welding material to assemble the spheres.

We, at Douglas, had the dubious honor of developing a number of important data points with respect to the hazards of liquid hydrogen. Personnel protection and the placement of test facilities always considered the total amount of propellant within a test vehicle that could mix and detonate. In 1960, we were told that, for LOX/LH2 mixtures, a factor of 65 percent was to be used—meaning that a mixture of 100,000 pounds of LOX/LH2 propellant would cause a blast equal to 65,000 pounds of TNT. This determined test stand location and facilities protection. The 1964 and 1967 explosions established two data points for the destruction of vehicles containing 100,000 pounds and 200,000 pounds, respectively, of propellant. The results clearly showed that the blast equivalency was only five percent. We demonstrated that LH2 could be stored for extended periods of time in spherical, vacuum-jacketed storage tanks, and that transfer was best accomplished by allowing full flow rather than slow transfer—which caused uneven thermal stresses on the pipes and components. We demonstrated that gaseous hydrogen could be safely vented by bubbling the gases through a water pond with igniters above to burn the gases in a controlled manner. We showed that a flight stage using LOX/LH2 propellants could be safely parked in orbit and successfully restarted to complete its mission.

NASA's directive to static fire all flight stages for acceptance was a time-consuming and costly program. However, the program's launch record is undeniable. Not a single flight vehicle was lost because of the Saturn vehicle's performance.

—Don Brincka
Former Douglas Chief Test Conductor at SACTO

One

Building SACTO

Here is a 1960s aerial view of the Alpha test complex, looking south. On the left is the single-position Alpha 1 test stand, and on the right is the dual-position Alpha 2A/B test stand. In between the test stands is the control center, linked to the test stands by reinforced concrete tunnels. Each test stand had a spherical liquid hydrogen storage tank and two horizontal liquid oxygen tanks. (Ralph Allen/MSFC.)

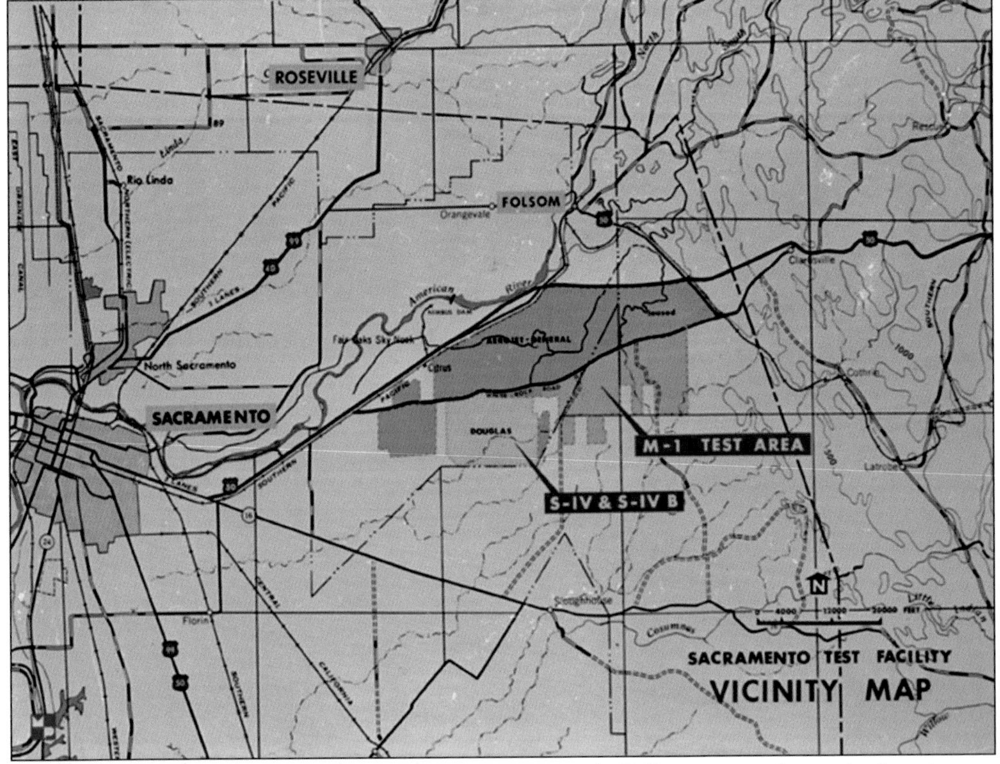

A 1960s-era map shows the location of the SACTO test facility, 15 miles east of Sacramento in Rancho Cordova. In 1950, Aerojet started to purchase former gold mining land, and by the mid-1950s, it had a 25,000-acre site. The Douglas Aircraft Company bought about 3,800 acres of the west side of this site for its own use. (Ralph Allen/MSFC.)

During the early 1960s, the Douglas Aircraft Company (DAC) placed extensive advertisements in prominent magazines and newspapers promoting its involvement in the space race. This advertisement appeared in *Time* magazine on November 30, 1962, and was accompanied by a second page explaining DAC's contribution to the Saturn rocket. This artist's rendering shows a test firing of the S-IV Battleship stage at SACTO's Test Stand 1. (Alan Lawrie.)

The Alpha complex was the first to be built. Constructed in 1957, it was initially used for the development and acceptance testing of Thor missiles, with the test stands simply being known as Test Stands 1 and 2A/B. It was converted for Saturn S-IV stages with the addition of liquid hydrogen propellant facilities and the capability to test at simulated altitude conditions. Test Stand 1 was converted by summer 1961 and Test Stand 2B by January 1963. (Ralph Allen/MSFC.)

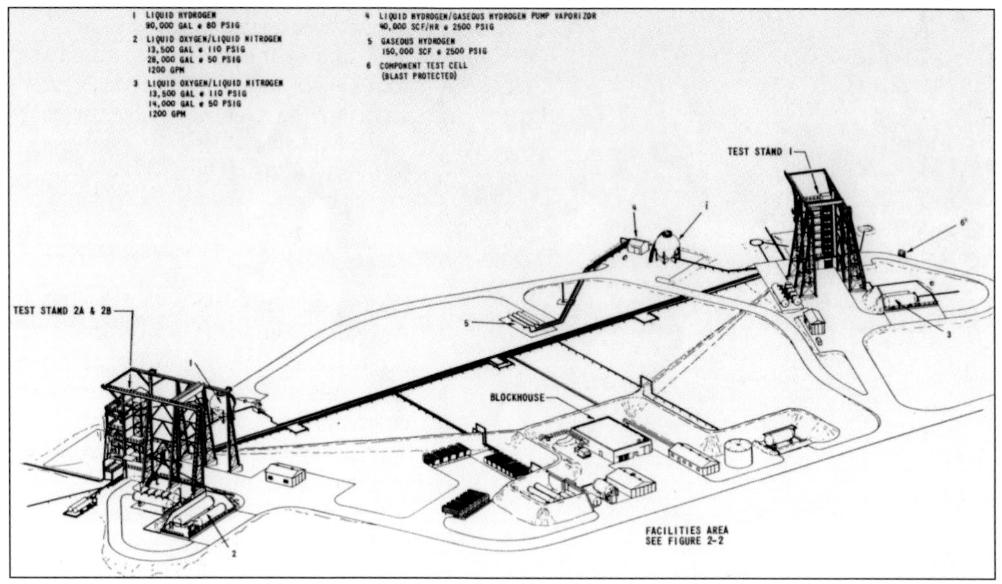

This diagram of the Alpha test complex shows Test Stand 1 (right) and Test Stand 2A/B (left), with the control center (blockhouse) between the test stands. The diagram shows the configuration after the conversion from Thor to S-IV testing. (Don Brincka.)

When Test Stands 1 and 2B were converted for the S-IV stage testing, each stand was provided with an insulated liquid hydrogen propellant tank. These tanks were built with an aluminum inner shell surrounded by a steel outer lining. The outer lining of the Test Stand 2 liquid hydrogen tank is shown being assembled on May 18, 1961. (Don Brincka.)

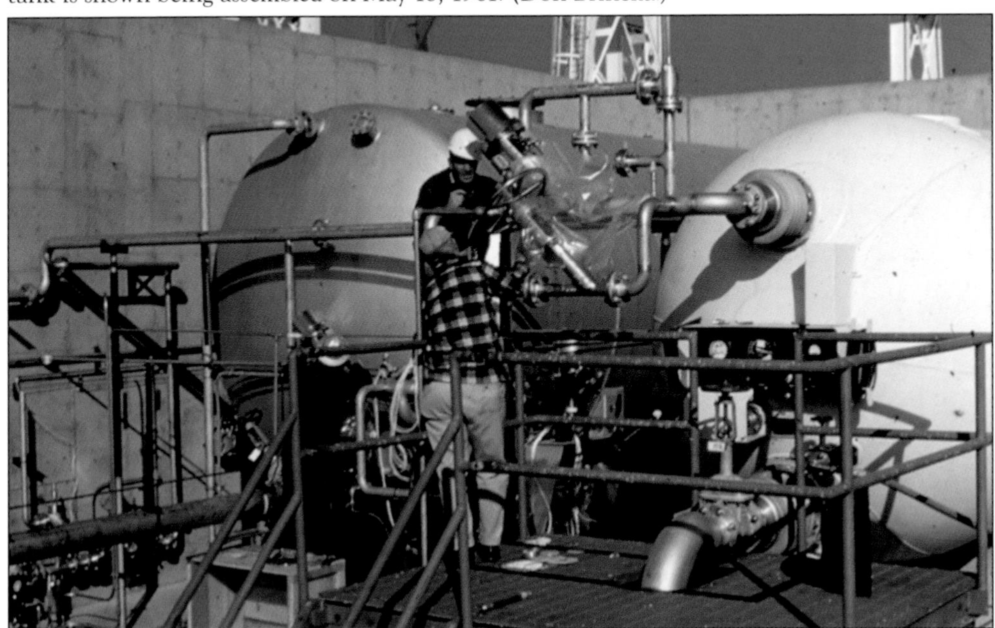

As liquid oxygen tanks were already in place at Test Stands 1 and 2A/B—to supply the Thor missiles—no conversion was needed for the S-IV stages. The tanks were located close to the test stands and were protected by a blast wall. Each test stand had two liquid oxygen tanks, as well as a liquid nitrogen tank for facility cryogenic checkout. The Test Stand 2 tanks are shown here. (Don Brincka.)

In order to simulate altitude pressure conditions at the outlet of the six RL-10 engines, 35-foot-long diffusers were attached to each engine—allowing a very low pressure to be maintained before and during firing. Each steel diffuser had a double wall to allow the flow of 3,100 gallons per minute of cooling water. The diffusers are shown being installed on Test Stand 2B in May 1962. (DAC 363234. Jim Porter/Terri Pennello.)

In the period between T-45 seconds and engine start, two-stage steam ejectors were used to achieve low pressure in the diffusers attached to each of the S-IV engines. Each stage of the ejectors was 30 feet long, and they were assembled in a vertical array on the front of the test stand. The ejectors are shown being installed on Test Stand 2B on May 29, 1962. (DAC 363300. Jim Porter/Terri Pennello.)

Pictured here is an artist's rendering of the Beta test complex. This early impression shows the third test stand (Beta II), which was never actually built. (Ralph Allen/MSFC.)

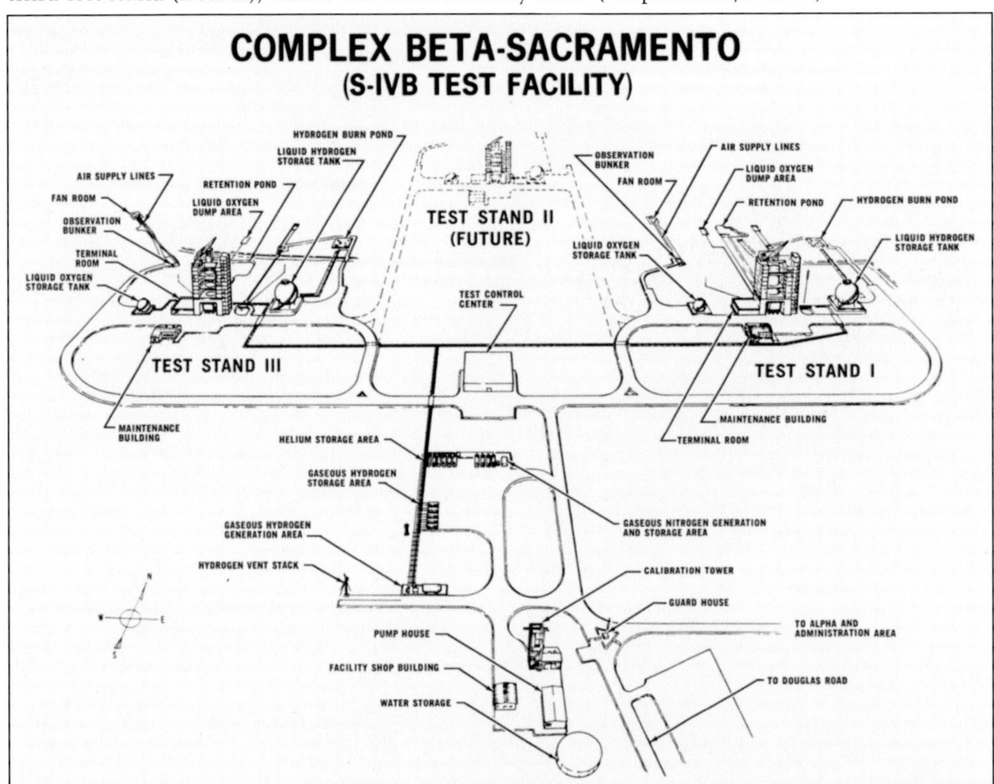

This diagram is of the Beta test complex. This illustration shows the planned, but never built, Beta II test stand. The Beta complex was built exclusively for testing the S-IVB-200 stages used on the Saturn IB launcher and the S-IVB-500 stages used on the Saturn V launcher. (Mike Jetzer.)

Construction of the Beta complex started in November 1962. The Paul Hardeman Company of Los Angeles was given the prime construction contract. By May 1963, when this photograph was taken, the Beta I test stand was taking shape, and the underground instrumentation tunnel running between the control center and the test stands was being installed. (DAC 389096. Jim Porter/Terri Pennello.)

A 1963 aerial view shows the construction of the Beta complex. In the foreground, work on the Beta III test stand was just beginning. In the background, the Beta I test stand was further along. Each test stand consisted of a reinforced concrete plinth with a steel gantry above. (Ralph Allen/MSFC.)

The opposite view shows the Beta I test stand in the foreground and the Beta III test stand in the background. The channels for the instrumentation tunnels had been prepared and the foundations laid for the control center between the two test stands. (Ralph Allen/MSFC.)

The construction of the Beta I test stand was well under way in July 1963. To the left of the test stand is the S-IVB Battleship. One of two S-IVB Battleships (the other one used at MSFC), this one was constructed by the Chicago Bridge and Iron Company on-site at SACTO, with Douglas adding the insulation. (DAC 394881. Jim Porter/Terri Pennello.)

This 1964 image shows the interior of the tunnel that ran from the Beta control center to the Beta I and Beta III test stands. The tunnel provided a secure passage for technicians as well as a protected link for instrumentation cables and services. (DAC 425250. Jim Porter/Terri Pennello.)

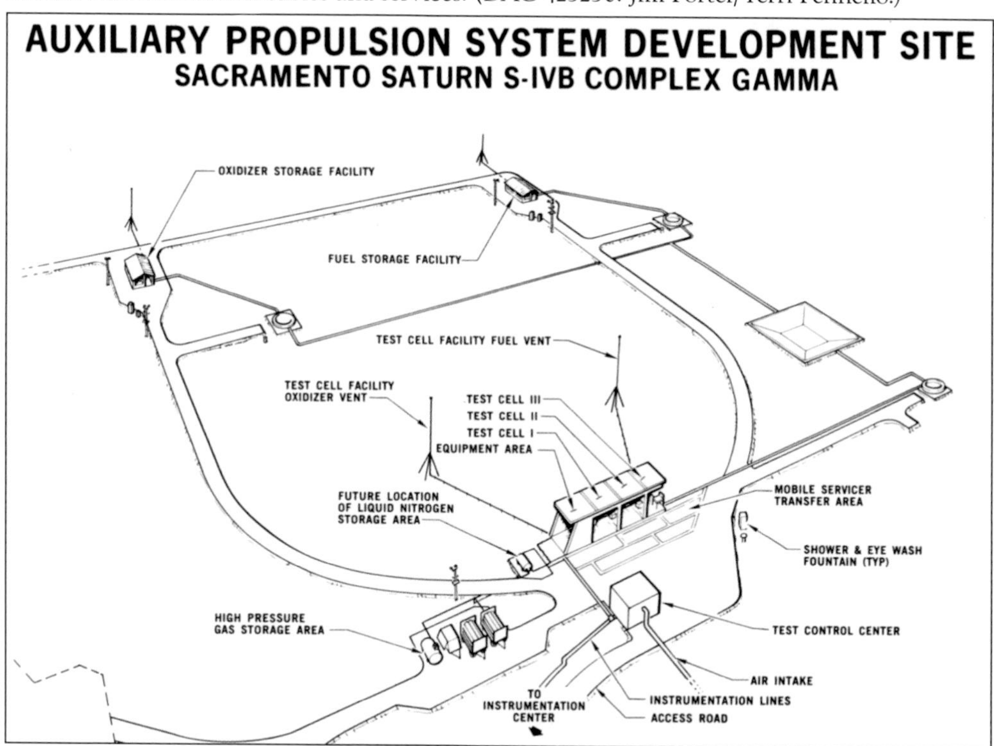

Here is a diagram of the Gamma test complex. This facility was built specifically to test fire the Auxiliary Propulsion Systems (APS) used on the S-IVB stages. These self-contained modules used monomethyl hydrazine and nitrogen tetroxide as propellants. Being hypergolic, the propellants react spontaneously without the need for a catalyst. (Mike Jetzer.)

This aerial photograph shows the Gamma test complex as the construction phase neared completion in July 1964. The Wismer-Becker Company performed construction work. Because of the reactivity of the propellants, they were stored in separate bunkers some way from each other and from the test stand. Gamma had its own control center. (Don Brincka.)

This aerial view of the whole SACTO site was taken after the Beta complex was built but before the Vehicle Checkout Laboratory was constructed. In the foreground is the administration area, with the Alpha test stands to the right and the Beta test stands in the distance to the left. The Gamma and Kappa test areas are behind the administration area. (Ralph Allen/MSFC.)

A close-up view of the Administration building shows the Beta test stands in the background to the left and the Alpha 2 test stand at the right. This is where the Douglas engineers and management had their offices. Residents from NASA and Rocketdyne also worked from here. (Ralph Allen/MSFC.)

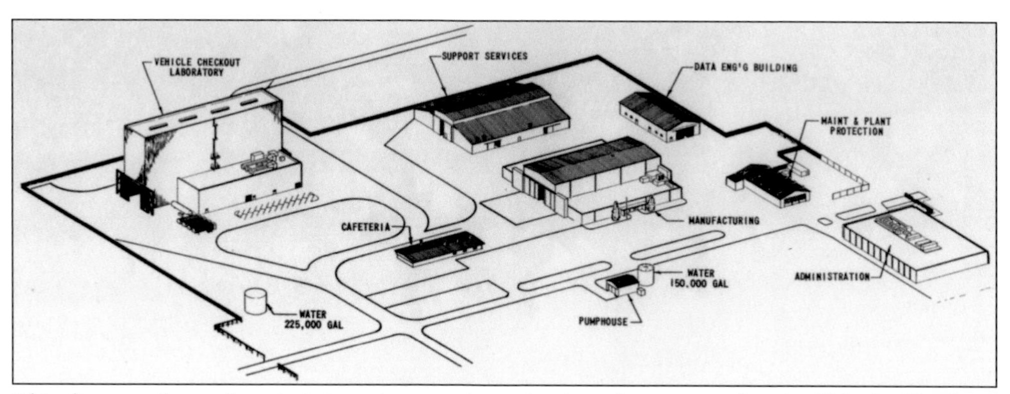

This diagram shows the administration area in its final configuration after the Vehicle Checkout Laboratory (VCL) was built. The VCL was used for checkout of the S-IVB stages, while the S-IV stages underwent checkout in the Support Services building (formerly known as the Evaluation and Development building) prior to the construction of the VCL. (Don Brincka.)

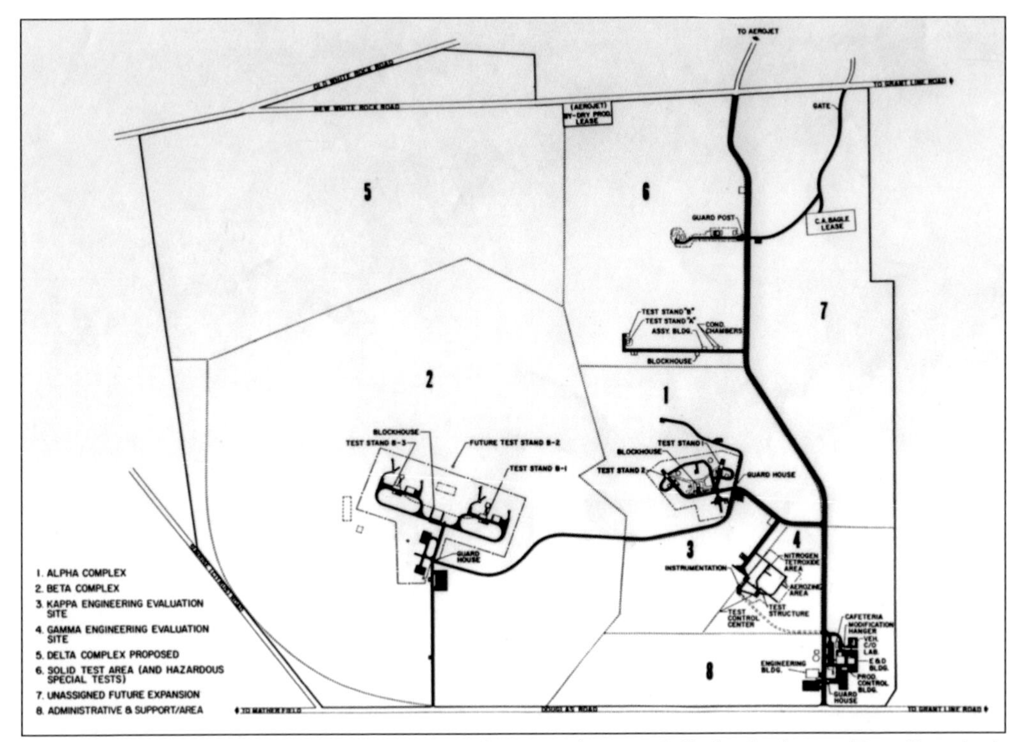

This is a map of the complete SACTO site after completion of all the facilities. (Don Brincka.)

Here is a 2009 satellite image of the SACTO site showing the location of the various test stands and facilities. Noticeable in this image are the parallel tailings that formed when the area was mined for gold. (Rebecca Allen.)

Two

Early Testing

Pictured here is a test firing of the S-IV Battleship in Test Stand 1. The photograph was taken between August 1962 and May 1963, when the Battleship firing program took place. Six Pratt & Whitney RL-10 engines, burning liquid oxygen and liquid hydrogen propellants, powered the S-IV stage. The S-IV Battleship was the only Saturn stage to be test fired in Test Stand 1. (Don Brincka.)

This rare photograph shows the test firing of a Thor intermediate-range ballistic missile (IRBM) in Test Stand 1. Around 10 Thor missiles were static fired at SACTO between 1958 and 1960 before being shipped to England for operational use by the Royal Air Force in East Anglia. The Thor used RP-1 kerosene and liquid oxygen as propellants. (Don Brincka.)

This September 1958 photograph shows the Douglas engineering group. This group photograph was taken so that copies could be sold to generate income for the families of the two Douglas employees killed in the July 30, 1958, Thor explosion at SACTO. (Val Sushkoff.)

A group photograph depicts the Blockhouse engineering crew. This September 1958 photograph was taken in front of the (later named) Alpha site control center for the same reason as the previous photograph. When hydrogen was first used with the S-IV stages, technicians used straw brooms to sweep the loaded stages. If hydrogen leaks were present, the brooms would catch fire! (Val Sushkoff.)

The S-IV Battleship stage is en route from the Douglas Aircraft Company in Santa Monica, where it was built, to the Los Angeles port of San Pedro on August 16, 1961. This test stage was then towed on an open-deck barge to the Courtland Dock, on the Sacramento River, before completing the overland journey to SACTO, where it was installed in Test Stand 1 on December 11, 1961. (Jim Porter/Terri Pennello.)

Installation of the six RL-10 engines into the diffusers below the S-IV Battleship stage took place at Test Stand 1 in May 1962. The diffusers were attached to each engine with a flexible seal and served as a vacuum chamber to provide low ambient pressure (less than 0.9 psia) from 45 seconds before engine ignition to the end of the firing in order to simulate the low pressure at altitude. (DAC 363229. Jim Porter/Terri Pennello.)

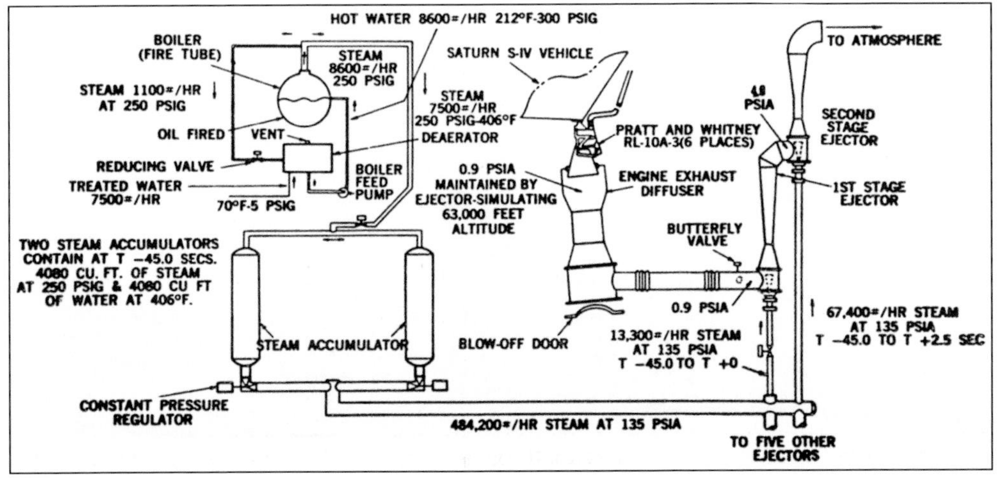

A diagram from a 1964 Douglas Missile and Space Systems Division conference paper explains the operation of the Alpha test site altitude system. Using diffusers below each engine and a two-stage steam ejector system, it was possible to simulate the low-pressure conditions at which the RL-10 engines of the S-IV stage would fire in space. (Mike Jetzer.)

Douglas technicians Gene Edwards (left) and Jerry Robison (right) install a helium heater on the S-IV Battleship, located in Test Stand 1, on July 16, 1962. The helium heater was a small engine fired up in space for the purpose of maintaining the liquid oxygen tank pressure as it dropped through the firing. (Jerry Robison.)

The S-IV Battleship undergoes one of its test firings in Test Stand 1 between August 1962 and May 1963. The S-IV used cryogenic propellants, which boiled off at ambient temperatures. Gaseous oxygen was vented out of the S-IV stage's vent and relief valve and can be seen at the top of the test stand, venting away. Excess hydrogen was piped away to a burn pond where it was combusted, thus eliminating the risk of an explosion close to the vehicle. (Gene Robinson.)

The S-IV Battleship was test fired 27 times between August 11, 1962, and May 4, 1963, with a cumulative firing time of 5,440.1 seconds. At the conclusion of this successful campaign, the RL-10 engines were removed, and the Battleship tank was hoisted out of Test Stand 1 on May 17, 1963—as shown in the photograph. (DAC 388654. Jim Porter/Terri Pennello.)

The scene inside the Alpha control center is captured during a critical test of an S-IV stage. While later testing at the Beta complex was largely automated, technicians performed most commands and instrumentation monitoring at the Alpha site. Douglas utilized the useful procedure of having technicians wear coats that identified their affiliation and responsibility in the test being conducted. (Gene Robinson.)

Each of the two Alpha test stands had an observation bunker, located quite close to and downrange of the test stand, so that technicians could observe the test firings in person. The Alpha bunkers were quite primitive and did not have protection on the backside. Technicians observed the firings through thick, protective glass windows. It appears that smoking was not prohibited! (Don Brincka.)

A group photograph was taken of the test stand crew in front of the Alpha 2A/B test stand. A Thor missile is located in the Alpha 2A location (right), and there is possibly an S-IV stage in the Alpha 2B location (left). The engine diffusers can be seen within the 2B gantry. This photograph was taken during a visit that Dr. Wernher von Braun made to SACTO. (Jerry Robison.)

A Californian National Guard tank is used as an up-range pillbox for observing the S-IV-5 stage acceptance firing in Test Stand 2B. The tank would provide protection for the observers in the event of an explosion. The S-IV-5 was the first flight stage to be test fired at SACTO. It was fired two times, on August 5 and 12, 1963, and this photograph was taken in June or July 1963. (DAC 394661. Jim Porter/Terri Pennello.)

Test Stand 2B was used for acceptance test firing of S-IV flight stages. On three occasions between May and December 1964 (May 7 and 8, August 27 and 28, and December 4 and 5), an S-IV stage was installed in the test stand within a day of the previous one being removed following testing. Such was the tight schedule that Douglas was working to. This image shows one of those occasions. (DAC 400473. Jim Porter/Terri Pennello.)

Three

Transportation

A variety of transportation methods was employed to move the various Saturn stages from the manufacturing factories in Southern California to the SACTO test site and then to Cape Kennedy (now Cape Canaveral) and other locations. Ocean-going ships, barges, trucks, and unique airplanes were used at various times. Here, the S-IVB-505N stage is shown being unloaded from the Super Guppy at Mather Air Force Base on August 17, 1967. (67-58398. Arlene Royer/NARA.)

On three occasions in the mid-1960s, before the Super Guppy aircraft became operational, ocean-going ships were used to transport S-IVB stages through the Panama Canal. The first part of the journey saw the stages trucked from SACTO to Courtland Dock, on the Sacramento River. Here, the S-IVB Battleship convoy is seen en route to Courtland Dock on November 3, 1965. (DAC A4583-668. Jim Porter/Terri Pennello.)

The S-IVB Battleship was the precursor for all the S-IVB stages tested at SACTO. At the completion of the SACTO testing, it was sent to the Arnold Engineering Development Center in Tennessee for environmental testing of the J-2 engine. For this trip to Courtland Dock, on November 3, 1965, it was necessary to trim trees to allow the convoy to pass. (DAC A4583-673. Jim Porter/Terri Pennello.)

Courtland Dock was the closest navigable water dock to SACTO. Towed barges were able to reach Courtland to collect and deliver rocket stages. The S-IVB Battleship is shown being lifted onto a derrick barge at Courtland Dock on November 3, 1965. The barge was towed to Mare Island Naval Shipyard, where the S-IVB Battleship was later transferred to a commercial ocean-going ship. (DAC A4583-681. Jim Porter/Terri Pennello.)

The Mare Island Naval Shipyard, near San Francisco, was capable of handling large ocean-going ships. The S-IVB Battleship is shown being transferred from the dock to the States Marine Lines vessel *Bayou State*, bound for New Orleans, on January 8, 1966. (DAC A4583-700. Jim Porter/Terri Pennello.)

The S-IVB-201 and S-IVB-202 were the only flight stages to be shipped from Mare Island to Cape Kennedy. While the S-IVB-202 stage employed NASA's ship *Point Barrow*, the S-IVB-201 stage was transported by the Isthmian Lines vessel *Steel Executive*. Loading at Mare Island took place on September 3, 1965. (DAC A4583-585. Jim Porter/Terri Pennello.)

The S-IVB stages were manufactured in Huntington Beach. Between February 1965 and April 1966, eight stages were shipped from Seal Beach Naval Weapons Station, located seven miles from Huntington Beach, to Courtland Dock on NASA's covered barge *Orion*. The S-IVB-204 stage is shown arriving at Courtland Dock on January 14, 1966, before being transported overland to SACTO for testing. (DAC A4583-776. Jim Porter/Terri Pennello.)

Between September 1963 and May 1965, NASA used its own specially built Pregnant Guppy aircraft to transport the S-IV flight stages from Santa Monica airport—next to the Douglas S-IV manufacturing plant—to Mather Air Force Base for testing at SACTO and onwards to Cape Kennedy after testing was completed. An S-IV stage is shown with the Pregnant Guppy at Mather Air Force Base. (Jim Porter/Terri Pennello.)

All six S-IV flight stages were transported aboard the Pregnant Guppy aircraft. To enable the stages to be loaded, the aircraft split into two parts. This image shows the final S-IV stage (S-IV-10) being loaded onto the Pregnant Guppy at Mather Air Force Base before the flight to Cape Kennedy on May 9, 1965. (Dick Serrano.)

Here, the S-IV-10 stage is secured within the Pregnant Guppy prior to attachment of the tail section. The aircraft took off from Mather Air Force Base for the flight to Cape Kennedy on May 9, 1965. On the flights to Cape Kennedy, up to three refueling stops were always needed. During a landing at Austin, Texas, with the S-IV-7 stage on board, the Pregnant Guppy blew both outboard tires. (Dick Serrano.)

Douglas manufactured the S-IVB stages at Huntington Beach. When complete, they were taken five miles to the Los Alamitos Naval Air Station and then flown to Mather Air Force Base in the new Super Guppy airplane for testing at SACTO. Here, midshipmen view the S-IVB-505N stage being loaded on the Super Guppy at Los Alamitos on August 17, 1967. (67-58386. Alan Lawrie/Arlene Royer/NARA.)

The Super Guppy replaced the Pregnant Guppy because the S-IVB stage had a wider diameter than the S-IV and would not fit in the Pregnant Guppy. From March 1966, the Super Guppy was the preferred means of transport. Here, the S-IVB-508 stage is seen being loaded in the Super Guppy at Los Alamitos Naval Air Station on December 30, 1968, for the flight to Mather Air Force Base. (69-72068. Alan Lawrie/Arlene Royer/NARA.)

The Super Guppy was operated for NASA by Aero Spacelines, and it was used for ferrying many large cargoes as well as Saturn rocket stages. It was purpose-built, using sections of several B-377 aircraft. The S-IVB-510 stage is shown here being loaded on the Super Guppy at Los Alamitos Naval Air Station on June 19, 1969, for the flight to Mather Air Force Base. (69-00082. Alan Lawrie/Arlene Royer/NARA.)

While the Pregnant Guppy's body split into two parts to allow access for the Saturn rocket, the front of the Super Guppy swung open on hinges. Here, the S-IVB-504N stage is being unloaded from the Super Guppy at Mather Air Force Base on June 16, 1967, after the flight from Los Alamitos Naval Air Station. (Phil Broad/Mike Jetzer.)

The S-IVB-211 stage is unloaded from the Super Guppy at Mather Air Force Base on October 18, 1968, before the road trip to SACTO. Due to a lack of missions, this stage was never tested and was placed in long-term storage at SACTO for two years. It never flew and can now be seen at the US Space and Rocket Center in Huntsville, Alabama. (68-70116. Arlene Royer/NARA.)

After testing at SACTO, the stages were taken by road back to Mather Air Force Base for the flight to Cape Kennedy. The photograph shows the S-IVB-502 stage waiting for the Super Guppy to land and pick it up for the trip to Kennedy Space Center on February 20, 1967. (Phil Broad/ Mike Jetzer.)

This photograph shows the loading of the S-IVB-205 stage on the Super Guppy at Mather Air Force Base on April 6, 1968, for the flight to Cape Kennedy. Because of delays in the Apollo program due to the Apollo 1 fire, this stage had been in storage at SACTO for two years after it was tested. It was launched on Apollo 7, the first manned Apollo flight, in October 1968. (68-66308. Phil Broad/Mike Jetzer.)

Mather Air Force Base was conveniently located just outside the SACTO perimeter, east of Sacramento. The Super Guppy was able to fly up the California coast without refueling and land at Mather Air Force Base with its Saturn cargo. Here, the S-IVB-504N stage is loaded on the Super Guppy at Mather Air Force Base on September 10, 1968, for the flight to Cape Kennedy. (68-70037. Alan Lawrie/Arlene Royer/NARA.)

The S-IVB-206 had been static fired at SACTO in 1966 before being delivered to Cape Kennedy, where it was stacked ready for launch. However, a fire on a parallel launch vehicle with the Apollo 1 capsule and crew meant that this stage had to be returned to SACTO for three years of storage. Here, the stage is being unloaded at Mather Air Force Base after returning from Cape Kennedy on April 14, 1967. (Phil Broad/Mike Jetzer.)

The Super Guppy takes off from Mather Air Force Base on the evening of December 2, 1968, with the S-IVB-505N stage bound for Kennedy Space Center. On a similar flight with the S-IVB-503N stage, on December 27, 1967, the Super Guppy crew had to make an emergency landing back at Mather Air Force Base after they heard loud popping noises. It transpired that one of the latches holding the Guppy closed was undone! (69-72000. Phil Broad/Mike Jetzer.)

The Super Guppy, with the S-IVB-506N stage onboard, climbs away from Mather Air Force Base on January 17, 1969, bound for Cape Kennedy. This is a historic image (which has never been published), as this was the stage that propelled the crew of Apollo 11 to the first moon landing in July 1969. (69-72078. Arlene Royer/NARA.)

Stage	Departure		Arrival		Transport
S-IV Battleship	11.8.61.	San Pedro	8.61.	Courtland Dock	Open deck barge
S-IV Battleship tank	21.5.63.	Courtland Dock	7.63.	New Orleans	
S-IV ASV	1.2.63.	San Pedro		Courtland Dock	Covered barge
S-IV-5	15.4.63.	San Pedro		Mare Island Naval Shipyard	Covered barge
S-IV-5		Mare Island Naval Shipyard	21.4.63.	Courtland Dock	
S-IV-5	20.9.63.	Mather Air Force Base	21.9.63.	Cape Canaveral	Pregnant Guppy
S-IV-6	27.9.63.	Santa Monica Airport	27.9.63.	Mather Air Force Base	Pregnant Guppy
S-IV-6	21.2.64.	Mather Air Force Base	22.2.64.	Cape Kennedy	Pregnant Guppy
S-IV-7	13.2.64.	Santa Monica Airport	13.2.64.	Mather Air Force Base	Pregnant Guppy
S-IV-7	10.6.64.	Mather Air Force Base	12.6.64.	Cape Kennedy	Pregnant Guppy
S-IV-8	7.8.64.	Santa Monica Airport	7.8.64.	Mather Air Force Base	Pregnant Guppy
S-IV-8	23.2.65.	Mather Air Force Base	26.2.65.	Cape Kennedy	Pregnant Guppy
S-IV-9	8.5.64.	Santa Monica Airport	8.5.64.	Mather Air Force Base	Pregnant Guppy
S-IV-9	21.10.64.	Mather Air Force Base	22.10.64.	Cape Kennedy	Pregnant Guppy
S-IV-10		Santa Monica Airport	5.11.64.	Mather Air Force Base	Pregnant Guppy
S-IV-10		Mather Air Force Base	10.5.65.	Cape Kennedy	Pregnant Guppy
S-IVB Common Bulkhead Test Article	9.65.	Seal Beach Naval Weapons Station	9.65.	Courtland Dock	Orion
S-IVB SACTO Battleship	3.11.65.	Courtland Dock	3.11.65.	Mare Island Naval Shipyard	Open deck barge
S-IVB SACTO Battleship	8.1.66.	Mare Island Naval Shipyard		New Orleans	Bayou State
S-IVB-F	12.2.65.	Seal Beach Naval Weapons Station	17.2.65.	Courtland Dock	Orion
S-IVB-F	10.6.65.	Courtland Dock	13.6.65.	Seal Beach Naval Weapons Station	Orion
S-IVB-201	30.4.65.	Seal Beach Naval Weapons Station	5.5.65.	Courtland Dock	Orion
S-IVB-201	3.9.65.	Courtland Dock		Mare Island Naval Shipyard	Open deck barge
S-IVB-201		Mare Island Naval Shipyard	19.9.65.	Cape Kennedy	Steel Executive
S-IVB-202	28.8.65.	Seal Beach Naval Weapons Station	1.9.65.	Courtland Dock	Orion
S-IVB-202	15.1.66.	Courtland Dock		Mare Island Naval Shipyard	Open deck barge
S-IVB-202 + aft inter-stage + S-IV mock-up		Mare Island Naval Shipyard	31.1.66.	Cape Kennedy	Point Barrow
S-IVB-203	29.10.65.	Seal Beach Naval Weapons Station	1.11.65.	Courtland Dock	Orion
S-IVB-203	4.4.66.	Mather Air Force Base	6.4.66.	Cape Kennedy	Super Guppy
S-IVB-204	10.1.66.	Seal Beach Naval Weapons Station	14.1.66.	Courtland Dock	Orion
S-IVB-204	6.8.66.	Mather Air Force Base	6.8.66.	Cape Kennedy	Super Guppy
S-IVB-205	9.4.66.	Seal Beach Naval Weapons Station	13.4.66.	Courtland Dock	Orion
S-IVB-205	6.4.68.	Mather Air Force Base	8.4.68.	Cape Kennedy	Super Guppy
S-IVB-206	30.6.66.	Los Alamitos Naval Air Station	1.7.66.	Mather Air Force Base	Super Guppy
S-IVB-206	13.12.66.	Mather Air Force Base	14.12.66.	Cape Kennedy	Super Guppy
S-IVB-206	13.4.67.	Cape Kennedy	14.4.67.	Mather Air Force Base	Super Guppy
S-IVB-206	3.8.70.	Mather Air Force Base	3.8.70.	Los Alamitos Naval Air Station	Super Guppy
S-IVB-207	30.8.66.	Los Alamitos Naval Air Station	31.8.66.	Mather Air Force Base	Super Guppy
S-IVB-207	1.5.70.	Mather Air Force Base	1.5.70.	Los Alamitos Naval Air Station	Super Guppy
S-IVB-208	2.12.66.	Los Alamitos Naval Air Station	2.12.66.	Mather Air Force Base	Super Guppy
S-IVB-208	13.10.70.	Mather Air Force Base	13.10.70.	Los Alamitos Naval Air Station	Super Guppy
S-IVB-209	9.3.67.	Los Alamitos Naval Air Station	9.3.67.	Mather Air Force Base	Super Guppy
S-IVB-209	22.7.70.	Mather Air Force Base	22.7.70.	Los Alamitos Naval Air Station	Super Guppy
S-IVB-211	17.10.68.	Los Alamitos Naval Air Station	18.10.68.	Mather Air Force Base	Super Guppy
S-IVB-211	15.9.70.	Mather Air Force Base	15.9.70.	Los Alamitos Naval Air Station	Super Guppy
S-IVB-500ST	15.12.65.	Seal Beach Naval Weapons Station	21.12.65.	Courtland Dock	Barge
S-IVB-500ST	30.3.66.	Mather Air Force Base	1.4.66.	MSFC	Super Guppy
S-IVB-501	11.3.66.	Seal Beach Naval Weapons Station	15.3.66.	Courtland Dock	Orion
S-IVB-501	12.8.66.	Mather Air Force Base	14.8.66.	KSC	Super Guppy
S-IVB-502	1.6.66.	Los Alamitos Naval Air Station	2.6.66.	Mather Air Force Base	Super Guppy
S-IVB-502	20.2.67.	Mather Air Force Base	21.2.67.	KSC	Super Guppy
S-IVB-503	11.10.66.	Los Alamitos Naval Air Station	11.10.66.	Mather Air Force Base	Super Guppy
S-IVB-503N	24.1.67.	Los Alamitos Naval Air Station	25.1.67.	Mather Air Force Base	Super Guppy
S-IVB-503N	27.12.67.	Mather Air Force Base	27.12.67.	Mather Air Force Base	Super Guppy
S-IVB-503N	29.12.67.	Mather Air Force Base	30.12.67.	KSC	Super Guppy
S-IVB-504N	16.6.67.	Los Alamitos Naval Air Station	16.6.67.	Mather Air Force Base	Super Guppy
S-IVB-504N	10.9.68.	Mather Air Force Base	12.9.68.	KSC	Super Guppy
S-IVB-505N	17.8.67.	Los Alamitos Naval Air Station	17.8.67.	Mather Air Force Base	Super Guppy
S-IVB-505N	2.12.68.	Mather Air Force Base	3.12.68.	KSC	Super Guppy
S-IVB-506N	25.1.68.	Los Alamitos Naval Air Station	25.1.68.	Mather Air Force Base	Super Guppy
S-IVB-506N	17.1.69.	Mather Air Force Base	18.1.69.	KSC	Super Guppy
S-IVB-507	7.8.68.	Los Alamitos Naval Air Station	7.8.68.	Mather Air Force Base	Super Guppy
S-IVB-507	6.3.69.	Mather Air Force Base	10.3.69.	KSC	Super Guppy
S-IVB-508	30.12.68.	Los Alamitos Naval Air Station	30.12.68.	Mather Air Force Base	Super Guppy
S-IVB-508	12.6.69.	Mather Air Force Base	13.6.69.	KSC	Super Guppy
S-IVB-509	31.3.69.	Los Alamitos Naval Air Station	31.3.69.	Mather Air Force Base	Super Guppy
S-IVB-509	17.1.70.	Mather Air Force Base	20.1.70.	KSC	Super Guppy
S-IVB-510	19.6.69.	Los Alamitos Naval Air Station	19.6.69.	Mather Air Force Base	Super Guppy
S-IVB-510	11.6.70.	Mather Air Force Base	12.6.70.	KSC	Super Guppy
S-IVB-511	16.9.69.	Los Alamitos Naval Air Station	16.9.69.	Mather Air Force Base	Super Guppy
S-IVB-511	29.6.70.	Mather Air Force Base	1.7.70.	KSC	Super Guppy

This chart shows all the Saturn-related transportation events to and from SACTO. (Alan Lawrie.)

Four

S-IVB Testing

Throughout the second half of the 1960s, SACTO was a very busy place, with Saturn rockets coming and going and being test fired on a regular basis. Here, the S-IVB-205, 206, and 207 stages are outside the north side of the Vehicle Checkout Laboratory on September 28, 1967. (67-60458. Arlene Royer/NARA.)

An unidentified man stands at the entrance to the Beta complex soon after construction was completed. To the right is the control center, and to the left is the Beta III test stand. In April 1967, the Douglas Aircraft Company underwent a merger and became McDonnell Douglas. (Vince Wheelock.)

Every stage that flew into Mather Air Force Base had to be driven a short distance to SACTO. After going along Douglas Road, the convoy would turn left into SACTO. Here, the S-IVB-505N stage is passing the Administration building at SACTO after arriving on August 18, 1967. (67-58399. Alan Lawrie/Arlene Royer/NARA.)

This stage was originally named S-IVB-504 but was renamed S-IVB-503N after the loss of the original S-IVB-503 vehicle in the January 1967 explosion. The stage is seen in the Beta I test stand in February 1967 prior to its static firing on May 3, 1967. (Dick Serrano.)

The S-IVB-206 stage was hoisted onto the Beta III test stand on November 22, 1967. The crane was located at the 13th deck level and had a capability to lift 50 tons. The stage had been returned from Cape Kennedy following the Apollo 1 fire and was now to undergo a post-storage checkout in Beta III—the first time this test stand had been used since the January 1967 explosion. (67-61109. Arlene Royer/NARA.)

In the background, the S-IVB-505N stage is being lifted by the gantry crane into the Beta I test stand for pre-firing checkout on September 1, 1967. In the foreground, the S-IVB-504N stage is being transported to the Vehicle Checkout Laboratory after undergoing two firings during August. (67-59473. Alan Lawrie/Arlene Royer/NARA.)

The S-IVB-505N stage approaches the Beta III test stand for pre-shipment modifications on August 1, 1968, as viewed from the top of Beta I. The stage had already undergone a single static firing in the Beta I test stand during the previous October. When the VCL was full, it was not uncommon for the Beta test stands to be used to support the stages during checkout and modification. (68-69213. Alan Lawrie/Arlene Royer/NARA.)

A night view shows the S-IVB-506N stage being lifted into the Beta III test stand on January 26, 1968. It remained there until August 1968, during which time a single static firing took place. This was the first firing in the Beta III test stand since it was damaged in the S-IVB-503 explosion in January 1967. (68-62966. Alan Lawrie/Arlene Royer/NARA.)

On November 23, 1968, the S-IVB-506N stage was raised into the Beta III test stand a second time. On this occasion, it was undergoing modification and checkout, and it remained in place for two months. It finally left SACTO on January 17, 1969, for Cape Kennedy, where it formed part of the Apollo 11 Saturn V rocket. (Phil Broad/Mike Jetzer.)

After two months of post-firing checkout in the Beta I test stand, the S-IVB-504N stage was removed on March 11, 1968. The stage had spent July and August of the previous year in the same test stand, during which two static firings had been conducted. The first firing had been aborted after a false indication of fire on the J-2 engine. (68-64834. Alan Lawrie/Arlene Royer/NARA.)

The S-IVB-507 stage was removed from the Beta I test stand on October 30, 1968. Its single static firing had occurred on October 16, 1968. Afterwards, the stage was moved to the VCL for post-static firing modification and checkout. On the left is the camera observation mast. (Phil Broad/Mike Jetzer.)

The 448.4-second static firing of the S-IVB-505N stage, held in the Beta I test stand on October 12, 1967, is viewed from the top of Beta III. The VCL is in the background. Each stage underwent a single firing as a check of workmanship and to calibrate the propulsion system. On some occasions, a second firing was needed after problems were encountered on the first firing. (67-60461. Alan Lawrie/Arlene Royer/NARA.)

This close-up shows the J-2 engine, serial number J-2101, on the S-IVB-506N stage during the static firing in the Beta III test stand on July 17, 1968. The 230,000-pound-thrust engine, built by Rocketdyne, burned liquid oxygen and liquid hydrogen propellants. Because of the high thrust level, it was not feasible to fire at simulated altitude conditions as had been done with the S-IV stage. (68-68411. Alan Lawrie/Arlene Royer/NARA.)

The S-IVB-507 stage undergoes static firing in the Beta I test stand on October 16, 1968. This firing marked the final use of the Beta I test stand. The 433.2-second firing was preceded by simulated elements of a mission flight, including a helium repressurization of the stage and two firings of the oxygen/hydrogen burner—used to heat the helium pressurizing gas. (68-70106. Alan Lawrie/ Arlene Royer/NARA.)

Pictured here is a downrange view of the test firing of an unidentified S-IVB stage at SACTO. In flight, the Saturn V S-IVB third stages were fired two times. The first firing completed the lifting of the Saturn V and Apollo vehicles into Earth orbit. The second firing, after stage repressurization, sent Apollo to the moon in what was called the trans-lunar injection burn. (Dick Serrano.)

The S-IVB-508 stage undergoes a 457.0-second static firing in the Beta III test stand on February 20, 1969, as viewed from the top of Beta I. This stage flew on Apollo 13 and was the first S-IVB stage to be purposely crashed into the moon after the completion of its mission, in order to generate seismic data. (69-73064. Alan Lawrie/Arlene Royer/NARA.)

The S-IVB-509 stage is static fired in the Beta III test stand for 452.4 seconds on May 14, 1969. Note the car parked behind the observation bunker. This stage flew on Apollo 14 and was also crashed into the moon's surface. (69-75855. Alan Lawrie/Arlene Royer/NARA.)

On August 14, 1969, the S-IVB-510 stage was static fired in the Beta III test stand for 448.7 seconds. This stage flew on Apollo 15 in July 1971 and had a similar fate to the Apollo 13 and 14 S-IVB stages—ending up being crashed into the moon's surface. (69-01857. Alan Lawrie/Arlene Royer/NARA.)

This mid-1960s photograph shows activities within the air-conditioned Beta control room. A single control center managed activities on both the Beta I and Beta III test stands. (Dick Serrano.)

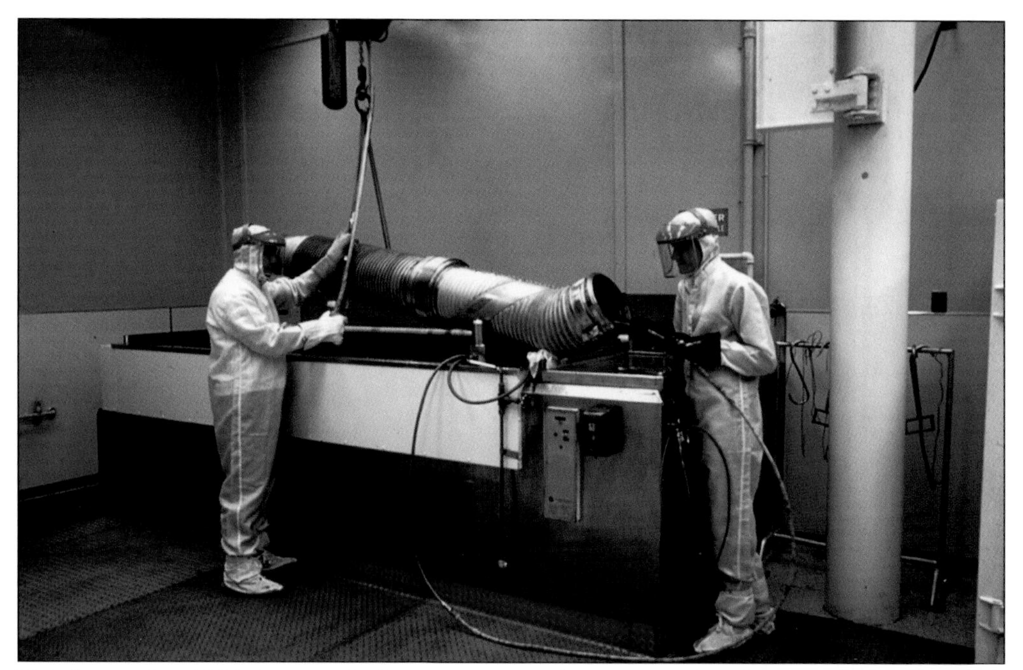

This September 1966 photograph shows activity in the clean room at SACTO. Technicians are degreasing a duct using trichloroethylene degreaser, after which the duct would be dried, inspected, rapped, and sent to the test stands. Earl Campbell is the technician on the left. (Dick Serrano.)

This image shows NASA and Douglas employees donating children's gifts to the Salvation Army outside the main administration building at SACTO in 1966. Julian Pennello, seen smoking a pipe, is the father of image supplier Terri Pennello. (DAC 456820. Julian Pennello.)

Dr. Wernher von Braun (center), head of the Marshall Space Flight Center in Huntsville, Alabama, made a number of visits to SACTO. In this photograph, taken in the Beta control center in 1964 or 1965, he is seen with Douglas officials Bill Duval, director of the Sacramento Test Center (left), and Jack Bromberg, vice president of the Missiles and Space Systems Division of the Douglas Aircraft Company (right). (A4583-144. Jim Porter/Terri Pennello.)

This image shows the storage, modification, and checkout of the S-IVB Auxiliary Propulsion System modules in the Maintenance and Assembly building during September 1968. There were two different designs of APS modules; the smaller ones were used on the Saturn IB launcher and the larger ones on the Saturn V. Each S-IVB stage had two modules attached, with each module containing its own complete propulsion system. (68-70038. Phil Broad/Mike Jetzer.)

A Saturn V–type S-IVB APS module is ready for firing at the Gamma complex test cell III in May 1968. Each module contained a number of small bipropellant thrusters and was used for attitude control of the S-IVB stage in flight, as well as ullage settling (though not on the Saturn IB variant), which forced the propellant to the tank outlet. Each module underwent a test firing before flight. (68-66896. Alan Lawrie/Arlene Royer/NARA.)

An APS module is shown being lifted out of the Gamma test cell III during September 1968, following its test firing. The hypergolic propellants, monomethyl hydrazine and nitrogen tetroxide, are very dangerous and required special care in handling. (68-70047. Phil Broad/Mike Jetzer.)

The two propellants for the Gamma complex were stored in two bunkers, separated from each other and from the test cells where the APS modules were fired. For an APS firing, mobile service carts were filled with the propellants from the storage bunkers and taken to the test cells. This activity is shown at the Gamma fuel storage area in May 1968. (Arlene Royer/NARA.)

Technicians pose in front of the Gamma test facility. A Saturn IB APS module, ready for hot fire testing, is located in test cell II. The Gamma complex propellant storage bunkers had the capacity to store 4,000 pounds of oxidizer and 750 pounds of fuel. (David Dallman.)

The Kappa test site was located between the Alpha and Gamma test sites. It was used for component and subsystem development and production acceptance testing. The photograph shows the test set-up at Kappa cell "A" for S-IVB fuel pre-valve testing in April 1968. (68-66309. Arlene Royer/NARA.)

Technicians remove an S-IVB helium tank from Kappa test cell "C" in September 1967. High pressure testing was performed in this protected bunker. (67-59472. Arlene Royer/NARA.)

In addition to being hot fire tested at the Gamma complex, APS modules underwent acceptance vibration testing at the Alpha 1 test site long after the site was no longer used for S-IV stage static firings. A small crane lifts a Saturn V APS module onto a handling fixture at the Alpha 1 site in September 1968. (68-70039. Phil Broad/Mike Jetzer.)

Another view shows an APS module being unloaded at the Alpha 1 site prior to being installed on the shaker unit in January 1969. By this time, the flame deflector plate had been removed from the Alpha 1 test stand, allowing a protected area for vibration testing to take place. (69-72081. Arlene Royer/NARA.)

A Saturn V APS module is on the vibration shaker, configured for testing in the tangential axis at Alpha 1, Site III, in September 1968. Vibration testing in all three axes was possible. Site III was the nomenclature given to the location of the vibration shaker within the Alpha 1 complex. (68-70043. Phil Broad/Mike Jetzer.)

Technicians install the liquid hydrogen/liquid oxygen burner at the edge of the S-IVB-505N stage thrust structure in Tower 7 at the McDonnell Douglas manufacturing plant in Huntington Beach in July 1967. The burners were used to heat up the helium in the stages' fuel and oxidizer tanks before the trans-lunar injection firing. They were test fired at SACTO while attached to the S-IVB stages in the Beta test stands. (Phil Broad/Mike Jetzer.)

After the last test firing at the Alpha complex in January 1965, the test site was used for a variety of other testing. In August 1967, an eight-foot-diameter tank with external insulation applied underwent cryogenic testing at the Alpha 1 test site to evaluate if the problematic foam insulation that was applied to the Saturn S-II second stage would bond to a very cold tank. (67-58395. Arlene Royer/NARA.)

The S-IVB-209 stage enters the door of the Evaluation and Development (later known as Support Services) building on July 19, 1967. It was to remain there in storage for three years. Earlier, this building was used to perform computerized checkout of horizontal S-IV stages, and therefore, it served the same function as the VCL would for the S-IVB stages. No propellants were used in this building. (Phil Broad/Mike Jetzer.)

The S-IVB-500ST stage arrives at the SACTO VCL Tower 1 (south bay) on December 22, 1965. The S-IVB-500ST stage was a third-stage breadboard simulator, used for developing integrated computer tapes. It underwent testing in the VCL during the first quarter of 1966 before being flown to MSFC in Huntsville in the first operational flight of the Super Guppy. (A4583-609. Jim Porter/Terri Pennello.)

The S-IVB-504N stage enters the south door of the VCL in June 1967 in preparation for pre-firing modification. The VCL was designed to perform simulated flights and conduct post-fire checks on stage subsystems. The VCL comprised two towers, a control room, and a maintenance support area. Site preparation for the VCL was started in October 1964, and actual construction of the building started two months later. (Dick Serrano.)

The S-IVB-206 stage is shown on November 12, 1967, being removed from Tower 2 (north bay) of the VCL, where it had been in storage for the previous seven months. It was moved to the Beta III test stand for post-storage checkout. A blast wall separated the north and south towers of the VCL in case there was an explosion of a stage during testing. (67-60559. Phil Broad/Mike Jetzer.)

In the foreground, the S-IVB-205 stage is supported on a roll dolly and technicians are entering the hydrogen tank via an airlock attached to the forward dome. Behind it, the S-IVB-207 stage is supported on the "birdcage," and in the background, the S-IVB-206 stage is located in the Tower 2 (north bay) gantry of the VCL. This photograph is from September 1967. (67-59476. Phil Broad/Mike Jetzer.)

In the background of this image from February 1967, the S-IVB-205 stage is in storage in the vertical gantry in Tower 2 (north bay) of the VCL. In the foreground, the S-IVB-207 stage is also in storage—lying horizontally. (Dick Serrano.)

The All Systems Test of the S-IVB-508 stage is monitored from the Vehicle Checkout control center at Huntington Beach in this May 1968 image. The stage was shipped to SACTO the following December. At SACTO, the VCL had a similar control center. It was located on the second floor and contained the necessary control consoles, computer, and related equipment to perform automatic and manual checkouts of S-IVB stages. (68-66887. Arlene Royer/NARA.)

The S-IVB-208 stage is horizontal and the S-IVB-209 stage is vertical in the south tower of the VCL. Based on this configuration, the photograph can be identified as being taken in March, April, May, or July 1967. (Phil Broad/Mike Jetzer.)

On the left, the S-IVB-206 stage sits outside the north side of the VCL. The S-IVB-207 and 205 stages are on the south side of the VCL in this photograph taken around March 7, 1968. The S-IVB-205 stage would fly on Apollo 7, in 1968; the S-IVB-206 stage on the first manned Skylab mission, in 1973; and the S-IVB-207 stage on the second manned Skylab flight, in 1973. (68-64836. Arlene Royer/NARA.)

Vertical and horizontal long-term storage covers are seen here during preliminary functional tests at Huntington Beach in December 1968. A number of stages went into long-term storage at SACTO and Huntington Beach. To ensure that cleanliness was maintained, the stages were covered in protective tents, and a positive pressure was applied to the interiors—ensuring that no particulates entered the tent. (69-72004. Phil Broad/Mike Jetzer.)

Following the completion of testing at SACTO, stages were airlifted to Cape Kennedy from Mather Air Force Base. First, they had to be transported by road from SACTO. Here, the S-IVB-505N stage is causing a traffic jam as it travels along Douglas Road and Mather Boulevard en route to Mather Air Force Base on December 2, 1968. In the background, SACTO's Alpha and Beta test stands are visible. (69-71996. Alan Lawrie/Arlene Royer/NARA.)

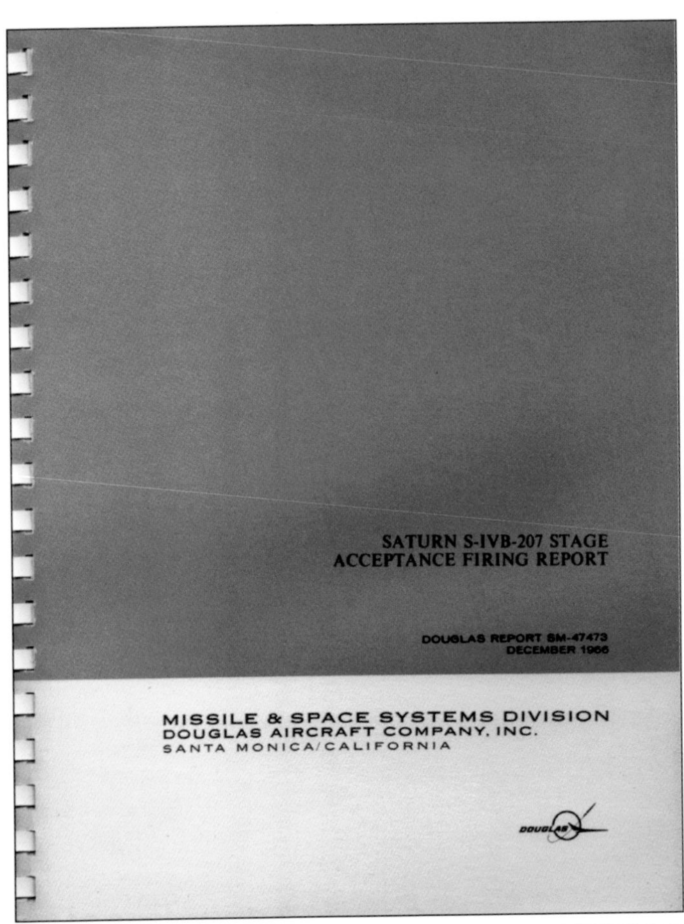
SATURN S-IVB-207 STAGE
ACCEPTANCE FIRING REPORT

DOUGLAS REPORT SM-47473
DECEMBER 1966

MISSILE & SPACE SYSTEMS DIVISION
DOUGLAS AIRCRAFT COMPANY, INC.
SANTA MONICA/CALIFORNIA

DOUGLAS

Every operation performed on the Saturn stages was controlled and documented by procedures and test reports. Shown here is the cover of the S-IVB-207 acceptance firing report—Douglas report SM-47473—dated December 1966. The report detailed all of the test parameters recorded during the firing. (Don Brincka/Alan Lawrie.)

The final test firing at the Beta I test stand took place on October 16, 1968, and at the Beta III test stand on December 18, 1969. This photograph, taken in September or October 1970, shows the Beta III test stand and the Beta control center during the interim phase—after testing had finished but before the site was decommissioned. (70-11876. Arlene Royer/NARA.)

A distant view shows the test firing of the S-IVB-506N stage, destined for flight as the Apollo 11 third stage. The firing took place in the Beta III test stand in the early evening of July 17, 1968. Also visible, on the left, is the Beta I test stand. (68-68408. Arlene Royer/NARA.)

This sunset view shows the S-IVB-507 stage in the Beta I test stand (left) and the S-IVB-505N stage in the Beta III test stand (right) in October 1968. The Saturn V rocket, as well as being the most powerful launch vehicle, had a 100-percent success rate. SACTO's contribution to this success was indeed immense. (68-70105. Arlene Royer/NARA.)

No.	Stage	Engine type	Test stand	Date	Duration
1	S-IV Battleship	RL10A-1	TS1	17.8.62.	10s
2	S-IV Battleship	RL10A-1	TS1	7.9.62.	13.6s
3	S-IV Battleship	RL10A-1	TS1	15.9.62.	28.3s
4	S-IV Battleship	RL10A-1	TS1	24.9.62.	60s
5	S-IV Battleship	RL10A-1	TS1	29.9.62.	42s
6	S-IV Battleship	RL10A-1	TS1	1.10.62.	7.2s
7	S-IV Battleship	RL10A-1	TS1	4.10.62.	420s
8	S-IV Battleship	RL10A-1	TS1	30.10.62.	70s
9	S-IV Battleship	RL10A-1	TS1	3.11.62.	448s
10	S-IV Battleship	RL10A-1	TS1	8.11.62.	38.5s
1	S-IV Battleship	RL10A-3	TS1	26.1.63.	468s
2-16	S-IV Battleship	RL10A-3	TS1	Various	Various
17	S-IV Battleship	RL10A-3	TS1	4.5.63.	444s
1	S-IV-5	RL10A-3	TS2B	5.8.63.	63.6s
2	S-IV-5	RL10A-3	TS2B	12.8.63.	476.4s
1	S-IV-6	RL10A-3	TS2B	22.11.63.	461s
1	S-IV-7	RL10A-3	TS2B	29.4.64.	485s
1	S-IV-8	RL10A-3	TS2B	20.11.64.	475.8s
1	S-IV-9	RL10A-3	TS2B	6.8.64.	398.94s
1	S-IV-10	RL10A-3	TS2B	21.1.65.	479.50s
1	S-IVB-200 Battleship	J-2	Beta I	1.12.64.	10.67s
2	S-IVB-200 Battleship	J-2	Beta I	9.12.64.	50.7s
3	S-IVB-200 Battleship	J-2	Beta I	15.12.64.	150.4s
4	S-IVB-200 Battleship	J-2	Beta I	23.12.64.	414.6s
5	S-IVB-200 Battleship	J-2	Beta I	13.3.65.	11.8s
6	S-IVB-200 Battleship	J-2	Beta I	19.3.65.	29.2s
7	S-IVB-200 Battleship	J-2	Beta I	31.3.65.	470s
8	S-IVB-200 Battleship	J-2	Beta I	7.4.65.	42s
9	S-IVB-200 Battleship	J-2	Beta I	15.4.65.	506.75s
10	S-IVB-200 Battleship	J-2	Beta I	27.4.65.	374s
11	S-IVB-200 Battleship	J-2	Beta I	4.5.65.	493.5s
1	S-IVB-201	J-2	Beta III	31.7.65.	0s
2	S-IVB-201	J-2	Beta III	8.8.65.	452s
1	S-IVB-202	J-2	Beta III	2.11.65.	0.41s
2	S-IVB-202	J-2	Beta III	9.11.65.	307s
3	S-IVB-202	J-2	Beta III	1.12.65.	463.8s
1	S-IVB-203	J-2	Beta I	26.2.66.	284.9s
1	S-IVB-204	J-2	Beta III	18.3.66.	451.2s
1	S-IVB-205	J-2	Beta III	2.6.66.	437.5s
1	S-IVB-206	J-2	Beta III	19.8.66.	433.7s
2	S-IVB-206	J-2	Beta III	14.9.66.	66.6s
1	S-IVB-207	J-2	Beta I	19.10.66.	445.6s
1	S-IVB-208	J-2	Beta I	12.1.67.	426.6s
1	S-IVB-209	J-2	Beta I	20.6.67.	455.95s
1	S-IVB-500 Battleship	J-2	Beta I	19.6.65.	8.92s
2	S-IVB-500 Battleship	J-2	Beta I	26.6.65.	167s
3	S-IVB-500 Battleship	J-2	Beta I	26.6.65.	3.84s
4	S-IVB-500 Battleship	J-2	Beta I	1.7.65.	5.45s
5	S-IVB-500 Battleship	J-2	Beta I	1.7.65.	1.72s
6	S-IVB-500 Battleship	J-2	Beta I	13.8.65.	16s
7	S-IVB-500 Battleship	J-2	Beta I	17.8.65.	170s
8	S-IVB-500 Battleship	J-2	Beta I	17.8.65.	319s
9	S-IVB-500 Battleship	J-2	Beta I	20.8.65.	170.9s
10	S-IVB-500 Battleship	J-2	Beta I	20.8.65.	360.2s
1	S-IVB-501	J-2	Beta I	20.5.66.	50s
2	S-IVB-501	J-2	Beta I	26.5.66.	151s
3	S-IVB-501	J-2	Beta I	26.5.66.	301s
1	S-IVB-502	J-2	Beta I	28.7.66.	150.7s
2	S-IVB-502	J-2	Beta I	28.7.66.	291.2s
1	S-IVB-503N	J-2	Beta I	3.5.67.	446.9s
1	S-IVB-504N	J-2	Beta I	23.8.67.	51.23s
2	S-IVB-504N	J-2	Beta I	26.8.67.	438s
1	S-IVB-505N	J-2	Beta I	12.10.67.	448.4s
1	S-IVB-506N	J-2	Beta III	17.7.68.	445.2s
1	S-IVB-507	J-2	Beta I	16.10.68.	433.2s
1	S-IVB-508	J-2	Beta III	20.2.69.	457.0s
1	S-IVB-509	J-2	Beta III	14.5.69.	452.4s
1	S-IVB-510	J-2	Beta III	14.8.69.	448.7s
1	S-IVB-511	J-2	Beta III	18.12.69.	442.8s

This chart details all of the Saturn-related test firings performed at SACTO. (Alan Lawrie.)

Five

Explosions

The SACTO site had a good safety record, due to the diligent efforts of the Douglas workforce and the management of NASA's Marshall Space Flight Center. However, three accidents did occur—in 1958, 1964, and 1967—which, it can be argued, were caused by a combination of schedule pressure and the use of new technologies. This image shows the destruction of the S-IVB-503 stage in January 1967. (G83-110. Jim Porter/Terri Pennello.)

In order to practice the procedures for quickly preparing a Thor missile for launch during the Cold War, the Initial Operational Capability (IOC) Site was established just south of Test Stand 1. Loading the propellant, rolling back the housing, and erecting the missile took just seven minutes. However, on July 30, 1958, an explosion took place at the location of the Thor seen at right, and two technicians were killed. (Val Sushkoff.)

The S-IV All Systems Vehicle (ASV) was a pathfinder for the flight stages, built with a flight-weight structure and full propulsion system. On January 24, 1964, preparations were underway for the first test firing at Test Stand 1. This still—taken from film of the test—shows observers in the observation bunker, watching during the final minute before planned ignition. (Mike Jetzer.)

The S-IV All Systems Vehicle exploded at Test Stand 1 on January 24, 1964, and was totally destroyed. The fireball reached a diameter of 380 feet, though nobody was injured in the accident. The root cause of the explosion was the S-IV ASV liquid oxygen tank, which was pressurized well beyond its design limit because the normal relief valve had stuck closed. (Mike Jetzer.)

In the aftermath of the S-IV ASV explosion, all the debris—as seen in this aerial photograph of Test Stand 1—was examined in detail. It was the first time that an explosion involving liquid oxygen and liquid hydrogen had been observed on the ground, and useful data on the extent of blast damage was obtained. (Don Brincka/Alan Lawrie.)

Another photograph shows the S-IV ASV debris around Test Stand 1 following the destruction of the stage. The investigation revealed that, after the relief valve froze closed, the pressure in the liquid oxygen tank was allowed to continue rising because there was no automatic warning in place, and pressure monitoring relied on an observer. This was a valuable lesson for the future. (Don Brincka/Alan Lawrie.)

The S-IVB-503 stage was due to undergo a normal flight acceptance test firing in the Beta III test stand on January 20, 1967. This stage would have been used on the Apollo 8 Saturn V, which was the first manned flight around the moon. At 511 seconds before the J-2 engine was due to start firing, the stage was totally destroyed when an explosion occurred, throwing debris in all directions. (Mike Jetzer.)

The film cameras located around the Beta III test stand had not yet been activated at the time of the S-IVB-503 explosion, so they missed the start of the event. A rapid manual activation resulted in both the previous film still and this one—captured moments later, as the fireball subsided. (Mike Jetzer.)

As the S-IVB-503 explosion occurred at dusk, the aerial survey started the following day and revealed the damage to the test stand and surrounding buildings. This was the only Saturn rocket stage destined for flight that was destroyed in a test. (G83-95. Jim Porter/Terri Pennello.)

This view is from the top of the Beta III test stand, looking down on a lower floor and the complete destruction of the S-IVB-503 stage. Remnants hang from the gantry. (DAC 18628. Jim Porter/Terri Pennello.)

The lower level of the Beta III test stand shows the remains of the S-IVB-503 stage's J-2 engine. This engine, serial number J-2061, had been delivered from Rocketdyne in April 1966 after undergoing four acceptance firings totaling 468.6 seconds. (DAC 18617. Jim Porter/Terri Pennello.)

The explosion of the S-IVB-503 stage blew sections off the roof of the nearby Butler building, which served as the test stand maintenance building. (Don Brincka/Alan Lawrie.)

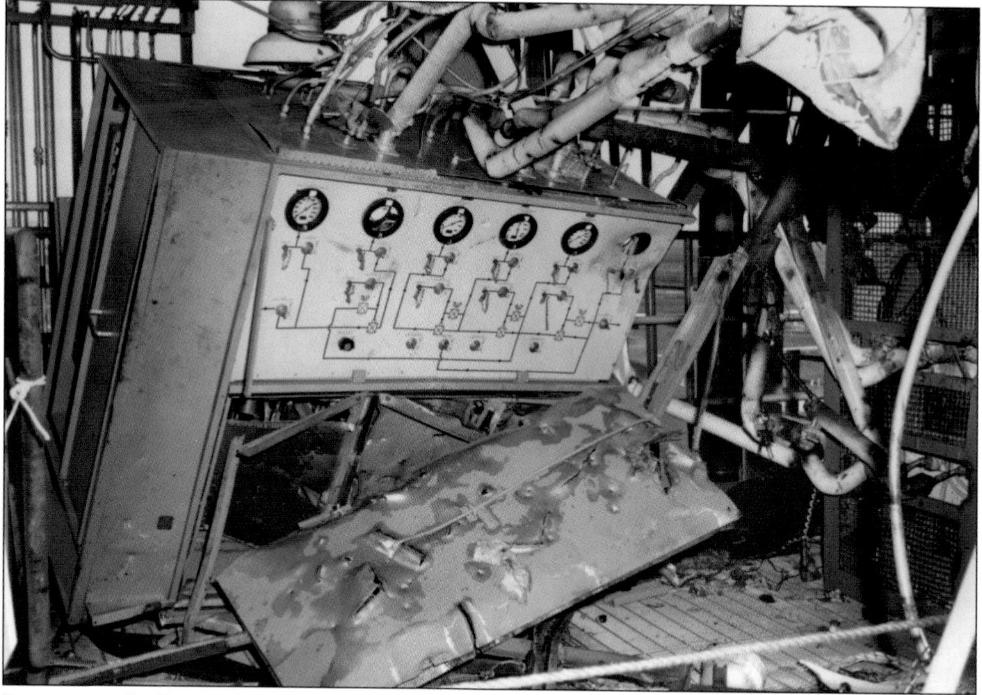

Damage to the Beta III test stand resulting from the explosion on January 20, 1967, was relatively minor. This was because the pressure wave from the blast passed through the steel framework and lattice floors without causing any buckling. However, solid items located on the Beta III test stand—such as the pictured Pneumatic Console B—fared poorly. (Don Brincka/Alan Lawrie.)

Following the explosion on January 20, 1967, a failure investigation team was formed to trace the cause of the accident. Crucial to this effort was the collection and examination of the debris from the S-IVB-503 stage, which was scattered in and around the test stand. Frogmen were dispatched into the ponds near the test stand to recover any fragments that had fallen into the water. (Don Brincka/Alan Lawrie.)

Located at the bottom of the S-IVB-503 stage, mounted to the thrust structure, were eight spherical, ambient-temperature helium storage tanks. Among the debris, five of these tanks were found intact. Only one hemisphere from each of the remaining three tanks was found. This photograph, taken in the concrete flame trench, shows a complete helium tank, as well as the lower hemisphere of the tank with the serial number 69. (Don Brincka/Alan Lawrie.)

The upper hemisphere of another helium tank (serial number 66) was found on the Beta III test stand gantry's fourth-level stairway. The outer surface of this hemisphere was blackened, indicating that the heat of the explosion had affected the tank—causing its internal pressure to rise until it ruptured. In flight, the helium would have been used to repressurize the Saturn S-IVB stage. (Don Brincka/Alan Lawrie.)

The upper hemisphere of the final helium tank (serial number 65) was found 250 feet up-range of the test stand, where it had fallen on the ground. Like tank 66, this tank had a blackened outer appearance—leading the failure team to conclude that this tank also had failed as a result of the stage explosion and had not caused it. (Don Brincka/Alan Lawrie.)

The lower hemisphere from helium tank 69 was found in the concrete flame trench, just downrange of the Beta III test stand. Here, a member of the failure investigation team supports the tank in situ. Unlike the other two tank hemispheres that were found, this one had a shiny outer surface—indicating that the heat of the explosion had not caused it to rupture. (Don Brincka/ Alan Lawrie.)

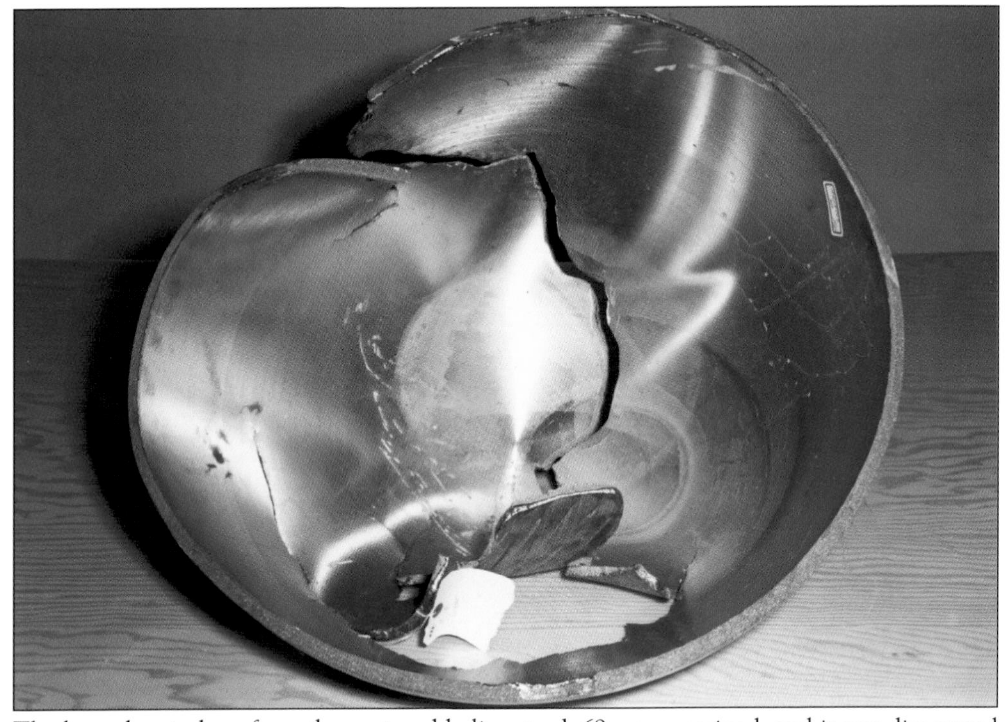

The lower hemisphere from the ruptured helium tank 69 was examined, and it was discovered that the upper and lower hemispheres had originally been welded together with the wrong grade of titanium filler wire. As a result, the weld had failed after several high pressure cycles—causing the upper tank hemisphere to penetrate the stage's propellant tanks and initiating the explosion. The upper hemisphere was never found. (DAC 18611. Jim Porter/Terri Pennello.)

Six

50 Years Later

In October 2006, the author was able to visit the remains of the SACTO site with former head of testing Don Brincka as his guide. The site has not seen any rocket testing since 1969, and in the late 1970s, the steel superstructure was removed from the test stand concrete plinths. These are the remains of the Alpha 1 test site—now used for grazing. (Alan Lawrie.)

The Beta I test stand also had been stripped of anything of value, leaving only the reinforced concrete pedestal. (Alan Lawrie.)

The Beta III test stand is seen here, with the remains of the liquid oxygen tank at the rear. In 1984, McDonnell Douglas sold the 3,800-acre SACTO site back to Aerojet—with the exception of the former administration area, which became known as Security Park and continues to be used. (Alan Lawrie.)

The author is pictured standing downrange of the remains of the Beta III test stand. (Alan Lawrie.)

In 2006, the Beta complex propellant storage tanks were largely intact. This is the Beta I test stand liquid hydrogen tank. (Alan Lawrie.)

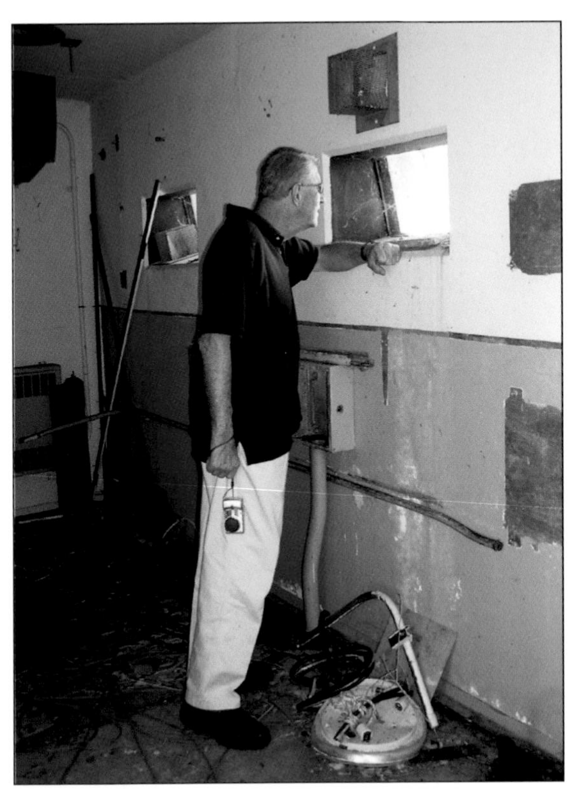

The observation bunkers were air-conditioned rooms, allowing technicians to spend lengthy periods monitoring stage activities and firings from a safe position protected by concrete walls and thick glass. Don Brincka looks through the window of the Beta III bunker in 2006. (Alan Lawrie.)

This photograph was taken from the Beta III test stand, looking back at the Beta I test stand. The remains of the control center are to the right, and the old VCL (now Security Park) is visible in the distance. (Alan Lawrie.)

In 2006, the structure of the Beta control center remained intact. However, anything of any value had long since been stripped from the interior. (Alan Lawrie.)

Several of the old site road signs and identification boards were still standing in 2006, albeit in a rather weather-battered state. The Beta control center is in the background. Compare this image with the period photograph at the top of page 44. (Alan Lawrie.)

The whole of the administration area is now known as Security Park and continues to be used by active businesses. The old VCL is occupied by Automotive Importing Manufacturing, Inc. Its chairman at that time, Frank Seabourne, showed the author around. This 2006 photograph shows that little had changed since the Saturn days. Compare this image with the period photograph at the top of page 62. (Alan Lawrie.)

Don Brincka poses in front of the old Administration building, just off Douglas Road at SACTO. From the outside, this building had changed little since the 1960s. Don showed the author his old office and the conference room where Dr. Wernher von Braun and other VIPs received their briefings on the progress of Saturn rocket testing in the 1960s. (Alan Lawrie.)

Since verifying that the underground water was free from contamination, clearance had been received to demolish the SACTO structures and construct housing. A team, comprised of (from left to right) Robert Hicks, Star AndersonHicks, Rebecca Allen, and Scott Baxter, as well as the author, performed a three-day survey to record the remaining buildings in March 2013. The team is seen in the Alpha 1 observation bunker, which may be compared with the photograph at the top of page 29. (Alan Lawrie.)

This is the view of the Alpha 1 test stand, as seen through the broken glass in the down-range observation bunker. This was the view that the observers would have had when the S-IV All Systems Vehicle exploded in January 1964—as seen in the photograph at the bottom of page 70. (Alan Lawrie.)

The team examines the IOC Site, south of the Alpha 1 test stand, where the Thor explosion took place in July 1958. The team recovered a sample of previously molten metal that presumably came from that explosion. (Alan Lawrie.)

The remains of the Alpha 1 test stand are seen with the concrete flame trench. The steel flame deflector plate had been removed in 1968, but the curved concrete supports remained. The red-and-white steel gantry had been removed in the 1970s. (Alan Lawrie.)

Rebecca Allen (left) and Scott Baxter remove one of several road signs found at the site. The signs were all in the standard Douglas shape and directed people to test stands and facilities not used in decades. In the background is the Alpha 1 test stand. (Alan Lawrie.)

Robert Hicks stands on top of a blast protection mound around the remains of the Alpha 2A/B test stand. Robert was the official photographer and returned to take a series of archival black-and-white photographs. (Alan Lawrie.)

A close-up shows the Alpha 2A/B test stand remains. Alpha 2A is closest to the camera. The team found a tunnel leading from the interior of the concrete structure shown. Square in cross section, it had concrete walls and led to the control center. A similar tunnel ran from the Alpha 1 site to the control center. (Alan Lawrie.)

This photograph shows the Alpha 2A (left) and Alpha 2B (right) test stands as seen from the concrete flame trench. (Alan Lawrie.)

Pictured is an exterior view of the Alpha control center. It was a reinforced building surrounded by a blast-protection mound. This was also the location of the period picture at the top of page 25. (Alan Lawrie.)

Inside the Alpha control center, all of the equipment and fittings had been removed some time ago. Compare this photograph with that at the bottom of page 28. (Alan Lawrie.)

Rebecca Allen (left) and Scott Baxter examine the inside of the bunker that was used to contain and protect the liquid oxygen storage tanks at the Alpha 1 test stand. The plinths to support the tanks were still in place. (Alan Lawrie.)

Katherine Anderson (center) provided additional support to the survey team and is seen here with Rebecca Allen (left) and the author (right) enjoying a picnic lunch on the roof of the terminal room attached to the Beta III test stand. A search with a metal detector located a section of metal later verified to be part of the S-IVB-503 stage, which exploded at this test stand in 1967. (Alan Lawrie.)

Prior to the 2013 survey, the Beta I test stand had already been demolished and many of the support structures at both Beta sites had been flattened—including the control center. All that remained of the Beta I test stand was the concrete foundation. This provided an opportunity to see the way the concrete structure had been reinforced with rebar rods. (Alan Lawrie.)

A close-up shows the two-inch-thick rebar used to reinforce the concrete found at the remains of the Beta I test stand. (Alan Lawrie.)

In 2013, the Gamma test site had not yet been demolished. In this end-on view, Cell III of the Gamma test stand is closest to the camera. In the foreground is one of several deluge showers located around the site, which were to be used by any personnel who came into contact with the dangerous hypergolic propellants used at the Gamma site. (Alan Lawrie.)

A side view of the Gamma test stand shows (from left to right) the equipment bay, Cell I, Cell II, and Cell III. At the far right is the Gamma control center, from which test firings of the APS modules could be observed and monitored. (Alan Lawrie.)

Pictured is the interior of the Gamma control center, as seen in 2013. All of the interior equipment had previously been removed. The protective glass windows, through which the tests were observed, were still in place. (Alan Lawrie.)

The southwest corner of the old Vehicle Checkout Laboratory is pictured in 2013. The building is still owned (as in 2006) by Automotive Importing Manufacturing, Inc., which is now under the chairmanship of Steve Seabourne, Frank's son. Steve took the team onto the higher gantry and into the old VCL control center adjacent to the test cells. (Alan Lawrie.)

This photograph shows the interior of the old VCL control center. Reinforced windows allowed technicians to view the systems testing being performed next-door on the S-IVB stages. It was noticeable that the doors were some feet above the current ground level. This was because, in the Saturn days, there was a false floor under which electrical cables and services ran. (Alan Lawrie.)

The author is seen at the high pressure Kappa cell "C" test bunker in 2013. Compare this with the period photograph at the bottom of page 57. The team completed their recording of the SACTO site in 2013, clearing the way for the final demolition of the test stands—to be replaced by housing. Perhaps, one day, a resident will find one of the three missing titanium hemispheres from the 1967 explosion when they dig their back garden. Information boards showing the history of SACTO have been erected at the site. (Alan Lawrie.)

Bibliography

Brincka, D.R. *Sacramento Test Center—Resources Handbook*. Douglas report SM 37538 R1. December 1966.

Hofferth, D.D., E.L. Wilson, and A.L. Polansky. "Altitude Simulation in Saturn S-IV Stage Testing." Douglas paper 3172. 1964.

Lawrie, Alan. "Return to Sacramento: A Review of Saturn Rocket Firings and Explosions." *AIAA 2007-5343*, July 2007.

———. *Saturn I/IB: The Complete Manufacturing and Test Records*. Burlington, ON: Apogee Books, 2008.

———. "The Saturn V Rocket: A New Review of Manufacturing, Testing and Logistics." *AIAA 2006-5031*, July 2006.

Lawrie, Alan, with Robert Godwin. *Saturn V: The Complete Manufacturing and Test Records*. Burlington, ON: Apogee Books, 2005.

Prentice, R.W. "Transportation of Douglas Saturn S-IVB Stages." Douglas paper 3688. November 1965.